Solutions Manual to Accompany

Loss Models

Solutions Manual to Accompany

Loss Models
FROM DATA TO DECISIONS

Stuart A. Klugman
Harry H. Panjer
Gordon E. Willmot

A Wiley-Interscience Publication
JOHN WILEY & SONS, INC.
New York / Chichester / Weinheim / Brisbane / Singapore / Toronto

ISBN 0471-23885-6

Printed in the United States of America.

10 9 8 7 6 5 4 3 2 1

Contents

Preface

This manual contains the solutions to the exercises in *Loss Models: From Data to Decisions*. We have not provided solutions to the open-ended, data-driven, questions that appear after the case study at the end of each chapter. Indeed, many of these questions do not have a unique answer. For those who choose to work these problems, we are providing a unique opportunity. Write up your solution and mail it (e-mail would be best, either in Adobe PDF format or something convertible to PDF (such as Word or DVI)) to (address correct at time of publication): Stuart Klugman, College of Business and Public Administration, Drake University, 2507 University Avenue, Des Moines, IA 50311, USA, Stuart.Klugman@drake.edu.

We will make all acceptable entries (subject to editing by us) available at the John Wiley & Sons web site (www.wiley.com) where other material related to the text may be found.

Acknowledgements for help with solutions were stated in the text, but the contribution of Clive Keatinge in working such a large number of the exercises deserves a second mention. Thanks, Clive.

Of course, all remaining errors are the fault of the authors. We welcome any corrections or improvements you would like to suggest.

S. A. KLUGMAN
H. H. PANJER
G. E. WILLMOT

Des Moines, Iowa
Waterloo, Ontario

1

Chapter 2 Solutions

1.1 SECTION 2.2

2.1 (a) $\hat{\mu} = \sum x_i/35 = 204{,}900$. $\hat{\mu}_2' = \sum x_i^2/35 = 1.4134 \times 10^{11}$. $\hat{\sigma} = 325{,}807$. $\hat{\mu}_3' = 1.70087 \times 10^{17}$, $\hat{\mu}_3 = 9.62339 \times 10^{16}$, $\hat{c} = 325{,}807/204{,}900 = 1.590078$, $\hat{\gamma}_1 = 2.78257$.
(b) $\hat{\pi}_{0.5} = y_{18} = 59{,}917$. $\hat{\pi}_{0.75} = y_{27} = 227{,}338$. $\hat{\pi}_{0.9} = 0.6y_{32} + 0.4y_{33} = 627{,}622$. $\hat{\pi}_{0.95} = 0.8y_{34} + 0.2y_{35} = 1{,}018{,}705$.
(c) B has a binomial distribution with $n = 35$ and $p = 0.75$. From that distribution, $\Pr(22 \le B < 31) = 0.9227$ while $\Pr(23 \le B < 31) = 0.8834$. The CI is from Y_{22} to Y_{31} which is $(103{,}217{,}000, 513{,}586{,}000)$.
(d) $E_n(500{,}000) = [\sum_{j=1}^{30} y_j + 5(500{,}000)]/35 = 153{,}139$.
$E_n^{(2)}(500{,}000) = [\sum_{j=1}^{30} y_j^2 + 5(500{,}000)^2]/35 = 53{,}732{,}687{,}032$.

2.2 (a) Using (2.2) the estimated moments are $\hat{\mu} = 7{,}415$, $\hat{\mu}_2' = 3.50687 \times 10^9$, $\hat{\mu}_3' = 5.33050 \times 10^{15}$. Then $\hat{\sigma} = 58{,}753$, $\hat{\mu}_3 = 5.25331 \times 10^{15}$, $\hat{c} = 58{,}753/7{,}415 = 7.9233$, and $\hat{\gamma}_1 = 25.903$. The histogram-based estimator of μ was used because that is the only basis we have for estimating the higher moments. A method for preserving the mean is given in Exercise 2.3
(b) The percentiles are $\hat{\pi}_{0.5} = 428$, $\hat{\pi}_{0.75} = 1{,}879$, $\hat{\pi}_{0.9} = 8{,}061$, and $\hat{\pi}_{0.95} = 20{,}981$.
(c) Again, using the histogram based estimator, the answers are $5{,}624$ and $620{,}487{,}472$.

2.3 Let $c = c_{j-1}$ and $d = c_j$. The two equations to solve are

$$n_j/n = \int_c^d r + sx\,dx = r(d - c) + s(d^2 - c^2)/2$$

$$a_j = \int_c^d x(r + sc)dx/(n_j/n),\ n_j a_j/n = r(d^2 - c^2)/2 + s(d^3 - c^3)/3$$

The solutions are

$$r = \frac{n_j}{n(d - c)^3}[4(d^2 + cd + d^2) - 6a_j(d + c)]$$

$$s = \frac{n_j}{n(d - c)^3}[12a_j - 6(d + c)]$$

Then, for $c_{j-1} \le u \le c_j$

$$\tilde{E}_n(u) = \sum_{i=1}^{j-1} \frac{n_i a_i}{n} + \int_{c_{j-1}}^u x(r + sx)dx + \int_u^{c_j} u(r + sx)dx + u \sum_{i=j+1}^k \frac{n_i}{n}$$

$$= \sum_{i=1}^{i-1} \frac{n_i a_i}{n} + r\frac{u^2 - c_{j-1}^2}{2} + s\frac{u^3 - c_{j-1}^3}{3}$$

$$+ ur(c_j - u) + us\frac{c_j^2 - u^2}{2} + u \sum_{i=j+1}^k \frac{n_i}{n}$$

For the fire loss data, the values are $r = 4.25569 \times 10^{-8}$ and $s = -8.31675 \times 10^{-14}$. This produces $\tilde{E}_{16,536}(350,000) = 4,299.86 + 1,276.71 - 755.44 + 2,234.24 - 1,855.68 + 571.48 = 5,771.17$.

2.4 The *pdf* is $f(x) = 2x^{-3}$, $x \ge 1$. The mean is $\int_1^\infty 2x^{-2}dx = 2$. The median is the solution to $.5 = F(x) = 1 - x^{-2}$, which is 1.4142. The mode is the value where the *pdf* is highest. Because the *pdf* is strictly decreasing, the mode is at its smallest value, 1.

2.5 $\hat{\mu}_1' = [2(2,000) + 6(4,000) + 12(6,000) + 10(8,000)]/30 = 6,000$,
$\hat{\mu}_2 = [2(-4,000)^2 + 6(-2,000)^2 + 12(0)^2 + 10(2,000)^2]/30 = 3,200,000$,
$\hat{\mu}_3 = [2(-4,000)^3 + 6(-2,000)^3 + 12(0)^3 + 10(2,000)^3]/30 = -3,200,000,000$.
$\hat{\gamma}_1 = -3,200,000,000/(3,200,000)^{1.5} = -0.55902$.

2.6 A single claim has mean $8,000/(5/3) = 4,800$ and variance

$$2(8,000)^2/[(5/3)(2/3)] - 4,800^2 = 92,160,000.$$

The sum of 100 claims has mean 480,000 and variance 9,216,000,000 which is a standard deviation of 96,000. The probability of exceeding 600,000 is approximately

$$1 - \Phi[(600{,}000 - 480{,}000)/60{,}000] = 1 - \Phi(1.25) = 0.106.$$

2.7 This is a single parameter Pareto distribution (see Appendix A with parameters $\alpha = 2.5$ and $\theta = 1$. The moments are $\mu_1 = 2.5/1.5 = 5/3$ and $\mu_2 = 2.5/.5 - (5/3)^2 = 20/9$. The coefficient of variation is $\sqrt{20/9}/(5/3) = 0.89443$.

2.8 We need the $.6(21) = 12.6th$ smallest observation. It is $0.4(38)+0.6(39) = 38.6$.

2.9 The mean of the gamma distribution is $5(1{,}000) = 5{,}000$ and the variance is $5(1{,}000)^2 = 5{,}000{,}000$. For 100 independent claims the mean is 500,000 and the variance is 500,000,000 for a standard deviation of 22,360.68. The probability of total claims exceeding 525,000 is

$$1 - \Phi[(525{,}000 - 500{,}000)/22{,}360.68] = 1 - \Phi(1.118) = 0.13178.$$

2.10 (a) We need the $.75(21) = 15.75th$ smallest observation. It is $0.25(13) + 0.75(14) = 13.75$.
(b) The ogive connects the points $(0.5, 0)$, $(2.5, 0.35)$, $(8.5, 0.65)$, $(15.5, 0.85)$,, and $(29.5, 1)$.
(c) The histogram has height $.35/2 = 0.175$ on the interval $(0.5, 2.5)$, height $0.3/6 = 0.05$ on the interval $(2.5, 8.5)$, height $0.2/7 = 0.028571$ on the interval $(8.5, 15.1)$, and height $0.15/14 = 0.010714$ on the interval $(15.5, 29.5)$.

2.11 $\mu = 0.05(100) + 0.2(200) + 0.5(300) + 0.2(400) + 0.05(500) = 300$.
$\sigma^2 = 0.05(-200)^2 + 0.2(-100)^2 + 0.5(0)^2 + 0.2(100)^2 + 0.05(200)^2 = 8{,}000$.
$\mu_3 = 0.05(-200)^3 + 0.2(-100)^3 + 0.5(0)^3 + 0.2(100)^3 + 0.05(200)^3 = 0$.
$\mu_4 = 0.05(-200)^4 + 0.2(-100)^4 + 0.5(0)^4 + 0.2(100)^4 + 0.05(200)^4 = 200{,}000{,}000$.
Skewness is $\gamma_1 = \mu_3/\sigma^3 = 0$. Kurtosis is $\gamma_2 = \mu_4/\sigma^4 = 200{,}000{,}000/8{,}000^2 = 3.125$.

2.12 (a) $100 \pm 1.96(75)/\sqrt{50} = 100 \pm 21 = (79, 121)$.
(b) Using (2.8), $v(\mu) = Var(\bar{X}) = Var(X)/n = \mu^2/n.$ $v(\hat{\mu}) = \bar{X}^2/n.$ CI is $100 \pm 1.96\sqrt{100^2/50} = 100 \pm 28 = (72, 128)$. Using Example 2.8, $.95 = \Pr\left(-1.96 \le \frac{\bar{X}-\mu}{\mu/\sqrt{n}} \le 1.96\right)$, $|\bar{X} - \mu| \le 1.96\mu/\sqrt{n}$, $(\bar{X} - \mu)^2 \le 3.8416\mu^2/n$, $\bar{X}^2 - 2\bar{X}\mu + \mu^2 - 3.8416\mu^2/n \le 0$, The CI is the solutions, $(79, 137)$.

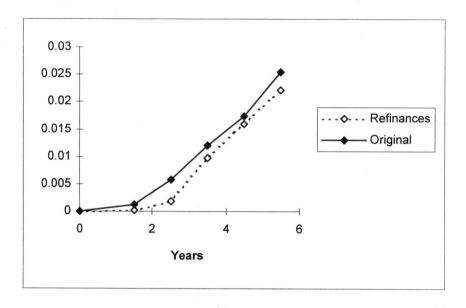

Fig. 1.1 Ogive for mortgage lifetime

2.13 The sum of 2,500 contracts has an approximate normal distribution with mean $2,500(1,300) = 3,250,000$ and standard deviation $\sqrt{2,500}(400) = 20,000$. The answer is $\Pr(X > 3,282,500) \doteq \Pr[Z > (3,282,500 - 3,250,000)/20,000] = \Pr(Z > 1.625) = 0.052$.

2.14 The plot appears in Figure 1.1. The points are the complements of the survival probabilities at the indicated times.

As one curve lies completely above the other it appears possible that original issues have a shorter lifetime.

2.15 The heights of the histogram bars are, respectively, $0.5/2 = 0.25$, $0.2/8 = 0.025$, $0.2/90 = 0.00222$, $0.1/900 = 0.000111$. The histogram appears in Figure 1.2.
(b) From the histogram the mean is estimated as $0.5(1) + 0.2(6) + 0.2(55) + 0.1(550) = 67.7$.

2.16 Following Example 2.10,

$$1 - \alpha = \Pr(a \leq B < b) = \Pr\left(\frac{a - 0.5 - np}{\sqrt{np(1 - p)}} \leq Z < \frac{b - 0.5 - np}{\sqrt{np(1 - p)}}\right)$$

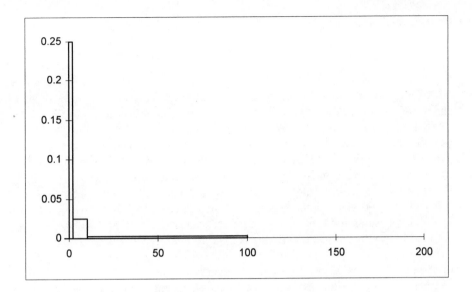

Fig. 1.2   Histogram for Exercise 2.15

Then $\frac{a-0.5-np}{\sqrt{np(1-p)}} = -z_{\alpha/2}$ and so $a = np + 0.5 - z_{\alpha/2}\sqrt{np(1-p)}$. Similarly, $b = np + 0.5 + z_{\alpha/2}\sqrt{np(1-p)}$.

2.17 $n = 5$, $p = 0.5$, $a = 2$, and $b = 4$. Then $1 - \alpha = \Pr(2 \leq B < 4)$ where $B \sim bin(5, 0.5)$. $1 - \alpha = \binom{5}{2}(0.5)^5 + \binom{5}{3}(0.5)^5 = 0.625$.

2.18 From the solution to Exercise 2.16,

$$1 - \alpha = \Pr\left(\frac{240 - 0.5 - 250}{\sqrt{125}} \leq Z < \frac{260 - 0.5 - 250}{\sqrt{125}}\right)$$
$$= \Pr(-0.939 < Z < 0.850) = 0.628.$$

1.2   SECTION 2.3

2.19 $E(\hat{\sigma}^2) = 5\sigma^2/6 = 10/6$. $bias = 10/6 - 2 = -1/3$.

2.20 $MSE = Var + bias^2$. $1 = Var + (0.2)^2$, $Var = 0.96$.

2.21 To be unbiased, $m = E(Z) = \alpha(0.8m) + \beta m = (0.8\alpha + \beta)m$ and so $1 = 0.8\alpha + \beta$ or $\beta = 1 - 0.8\alpha$. Then $Var(Z) = \alpha^2 m^2 + \beta^2 1.5m^2 = [\alpha^2 + (1 - $

$0.8\alpha)^2 1.5]m^2$ which is minimized when $\alpha^2 + 1.5 - 2.4\alpha + 0.96\alpha^2$ is minimized. This occurs when $3.92\alpha - 2.4 = 0$ or $\alpha = 0.6122$. Then $\beta = 1 - 0.8(0.6122) = 0.5102$.

2.22 One way to solve this problem is to list the 20 possible samples of size three and assign probability $1/20$ to each. The population mean is $(1 + 1 + 2 + 3 + 5 + 10)/6 = 11/3$.

(a) The twenty sample means have an average of $11/3$ and so the bias is zero. The variance of the sample means (dividing by 20 because this is the population of sample means) is 1.9778 and this is also the *MSE*.

(b) The twenty sample medians have an average of 2.7 and so the bias is $2.7 - 11/3 = -0.9667$. The variance is 1.81 and the *MSE* is 2.7444.

(c) The twenty sample midranges have an average of 4.15 and so the bias is $4.15 - 11/3 = 0.4833$. The variance is 2.65 and the *MSE* is 2.8861.

(d) $E(aX_{(1)} + bX_{(2)} + cX_{(3)}) = 1.25a + 2.7b + 7.05c$ where the expected values of the order statistics can be found by averaging the twenty values from the enumerated population. To be unbiased the expected value must be $11/3$ and so the restriction is $1.25a + 2.7b + 7.05c = 11/3$. With this restriction, the *MSE* is minimized at $a = 1.445337$, $b = 0.043733$, and $c = 0.247080$ with a *MSE* of 1.620325. With no restrictions, the minimum is at $a = 1.289870$, $b = 0.039029$, and $c = 0.220507$ with a *MSE* of 1.446047 (and a bias of -0.3944).

2.23 $bias(\hat{\theta}_1) = 165/75 - 2 = 0.2$, $Var(\hat{\theta}_1) = 375/75 - (165/75)^2 = 0.16$, $MSE(\hat{\theta}_1) = 0.16 + (0.2)^2 = 0.2$. $bias(\hat{\theta}_2) = 147/75 - 2 = -0.04$, $Var(\hat{\theta}_2) = 312/75 - (147/75)^2 = 0.3184$, $MSE(\hat{\theta}_2) = 0.3184 + (-0.04)^2 = 0.32$. The relative efficiency is $0.2/0.32 = 0.625$.

1.3   SECTION 2.4

2.24 (a)

$$
\begin{aligned}
G_j(\theta) &= E(X; c_j) - E(X; c_{j-1}) \\
&= \int_0^{c_j} x f(x) dx + c_j[1 - F(c_j)] \\
&\quad - \int_0^{c_{j-1}} x f(x) dx + c_{j-1}[1 - F(c_{j-1})] \\
&= \int_{c_{j-1}}^{c_j} x f(x) dx + c_j[1 - F(c_j)] - c_{j-1}[1 - F(c_{j-1})]
\end{aligned}
$$

(b) The expected value is

$$\int_0^{c_{j-1}} 0f(x)dx + \int_{c_{j-1}}^{c_j} (x - c_{j-1})f(x)dx + \int_{c_j}^{\infty} (c_j - c_{j-1})f(x)dx$$

$$= \int_0^{c_j} xf(x)dx - \int_0^{c_{j-1}} xf(x)dx - c_{j-1}[F(c_j) - F(c_{j-1})]$$
$$+ (c_j - c_{j-1})[1 - F(c_j)]$$

$$= \int_0^{c_j} xf(x)dx + c_j[1 - F(c_j)]$$
$$- \int_0^{c_{j-1}} xf(x)dx + c_{j-1}[1 - F(c_{j-1})]$$

$$= E(X; c_j) - E(X; c_{j-1})$$

2.25 $L = 2^n \theta^n (\Pi x_j) \exp(-\theta \Sigma x_j^2)$, $l = n \log 2 + n \log \theta + \Sigma \log x_j - \theta \Sigma x_j^2$, $l' = n\theta^{-1} - \Sigma x_j^2 = 0$, $\hat{\theta} = n/\Sigma x_j^2$.

2.26 $\hat{\mu} = 975$, $\hat{\mu}_2' = 977,916\frac{2}{3}$, $\hat{\sigma}^2 = 977,916\frac{2}{3} - 975^2 = 27,291\frac{2}{3}$. The moment equations are $975 = \alpha\theta$ and $27,291\frac{2}{3} = \alpha\theta^2$. The solutions are $\hat{\alpha} = 34.8321$ and $\hat{\theta} = 27.9915$.

2.27 $F(x) = (x/\theta)^\gamma / [1 + (x/\theta)^\gamma]$. The equations are $0.2 = (100/\theta)^\gamma / [1 + (100/\theta)^\gamma]$ and $0.8 = (400/\theta)^\gamma / [1 + (400/\theta)^\gamma]$. From the first equation $0.2 = 0.8(100/\theta)^\gamma$ or $\theta^\gamma = 4(100)^\gamma$. Insert this in the second equation to get $0.8 = 4^{\gamma-1}/(1 + 4^{\gamma-1})$ and so $\hat{\gamma} = 2$ and then $\hat{\theta} = 200$.

2.28 Table 1.1 provides the calculations.
(a) An example calculation for the sum is $(3,900 - 4,047)^2/3,900 = 6$ and the total is 158.
(b) The sum is minimized at $\alpha = 4.9283$ and $\theta = 134,675$.
(c) The *mle* is found by summing terms like $-3,900 \log[F(25,000) - F(0)] = 2,302$. The maximum occurs at $\alpha = 5.14825$ and $\theta = 147,130$.

2.29 (a) $f(x) = px^{p-1}$, $L = p^n (\Pi x_j)^{p-1}$, $l = n \log p + (p-1)\Sigma \log x_j$, $l' = np^{-1} + \Sigma \log x_j = 0$, $\hat{p} = -n/\Sigma \log x_j$.
(b) $E(X) = \int_0^1 px^p dx = p/(1+p) = \bar{x}$. $\hat{p} = \bar{x}/(1 - \bar{x})$.

2.30 (a) $\hat{\mu} = 3,800 = \alpha\theta$. $\mu_2' = 16,332,000$, $\hat{\sigma}^2 = 1,892,000 = \alpha\theta^2$. $\hat{\alpha} = 7.6321$, $\hat{\theta} = 497.89$.

Table 1.1 Calculations for Exercise 2.28

Range	n	G_j	$\alpha = 3$ $\theta = 75,000$ $G_j(\theta)$	χ^2	$\alpha = 4.9283$ $\theta = 134,675$ $G_j(\theta)$	χ^2	$\alpha = 5.14825$ $\theta = 147,130$ $-l$
0–25,000	7,000	3,900	4,047	6	3,976	2	2 ,302
25,000–50,000	7,000	1,500	1,441	2	1,548	2	2,244
50,000–60,000	2,000	200	89	62	97	53	600
60,000–100,000	2,000	300	186	44	196	36	683
50,000–	5,000	900	1,080	36	1,055	27	1,356
100,000–	2,000	200	157	9	130	25	534
Total				158		144	7,718

(b) $L = (\Pi x_j)^{\alpha-1} \exp(-\Sigma x_j/\theta)[\Gamma(\alpha)]^{-n}\theta^{-n\alpha}$.

$$
\begin{aligned}
l &= (\alpha-1)\Sigma \log x_j - \theta^{-1}\Sigma x_j - n\log\Gamma(\alpha) - n\alpha\log\theta \\
&= 81.61837(\alpha-1) - 38,000\theta^{-1} - 10\log\Gamma(\alpha) - 10\alpha\log\theta. \\
\partial l/\partial\theta &= 38,000\theta^{-2} - 10\alpha\theta^{-1} = 0,
\end{aligned}
$$

$\hat\theta = 38,000/(10 \cdot 12) = 316.67$.

(c) l is maximized at $\hat\alpha = 6.341$ and $\hat\theta = 599.3$ (with $l = -86.835$).

2.31 (a) $\hat\mu' = 2,000 = \exp(\mu + \sigma^2/2)$, $\hat\mu'_2 = 6,000,000 = \exp(2\mu + 2\sigma^2)$. $7.690090 = \mu+\sigma^2/2$ and $15.60727 = 2\mu+2\sigma^2$. The solutions are $\hat\mu = 7.39817$ and $\hat\sigma = 0.636761$.

$$
\begin{aligned}
\Pr(X > 4,500) &= 1 - \Phi[(\log 4,500 - 7.39817)/0.636761] \\
&= 1 - \Phi(1.5919) = 0.056.
\end{aligned}
$$

(b) $\hat\mu = \frac{1}{5}\Sigma \log x_i = 7.33429$. $\hat\sigma^2 = \frac{1}{5}\Sigma(\log x_i)^2 - 7.33429^2 = 0.567405$, $\hat\sigma = 0.753263$.

$$
\begin{aligned}
\Pr(X > 4,500) &= 1 - \Phi[(\log 4,500 - 7.33429)/0.753263] \\
&= 1 - \Phi(1.4305) = 0.076.
\end{aligned}
$$

2.32 $L = \theta^{-n} \exp(-\Sigma x_j/\theta)$. $l = -n\log\theta - \theta^{-1}\Sigma x_j$. $l' = -n\theta^{-1} + \theta^{-2}\Sigma x_j = 0$. $\hat\theta = \Sigma x_j/n$.

2.33 (a) $\hat\mu = 4.2 = (\beta/2)\sqrt{2\pi}$. $\hat\beta = 3.35112$.

(b) $L = \beta^{-10}(\Pi x_j)\exp[-\Sigma x_j^2/(2\beta^2)]$. $l = -10\log\beta + \Sigma\log x_j - \Sigma x_j^2/(2\beta^2)$.
$l' = -10\beta^{-1} + \beta^{-3}\Sigma x_j^2 = 0$. $\hat{\beta} = \sqrt{\Sigma x_j^2/10} = 3.20031$.

2.34 $f(x) = \alpha x^{-\alpha-1}$. $L = \alpha^n(\Pi x_j)^{-\alpha-1}$. $l = n\log\alpha - (\alpha+1)\Sigma\log x_j$.
$l' = n\alpha^{-1} - \Sigma\log x_j = 0$. $\hat{\alpha} = n/\Sigma\log x_j$.

2.35 (a) X is Pareto and so $E(X) = 1{,}000/(\alpha-1) = \bar{x} = 318.4$. $\hat{\alpha} = 4.141$.

(b)
$$L = \alpha^5\lambda^{5\alpha}[\Pi(\lambda+x_j)]^{-\alpha-1}$$
$$l = 5\log\alpha + 5\alpha\log 1{,}000 - (\alpha+1)\Sigma\log(1{,}000+x_j)$$
$$l' = 5\alpha^{-1} + 34.5388 - 35.8331 = 0. \ \hat{\alpha} = 3.8629.$$

2.36 $l = \sum_{j=1}^{J}\sum_{i=1}^{k_j} n_{ij}\log P_{ij}$ where P_{ij} is the probability of an observation being in the ith group of the jth sample. Then $\mathbf{S}(\boldsymbol{\theta})_r = \partial l/\partial\theta_r = \sum_{j=1}^{J}\sum_{i=1}^{k_j} n_{ij}P_{ij}^{-1}\partial P_{ij}/\partial\theta_r$. For the information term,

$$\partial^2 l/\partial\theta_s\partial\theta_r = \sum_{j=1}^{J}\sum_{i=1}^{k_j} n_{ij}\frac{P_{ij}\partial^2 P_{ij}/\partial\theta_s\partial\theta_r - (\partial P_{ij}/\partial\theta)(\partial P_{ij}/\partial\theta_s)}{P_{ij}^2}.$$

The quantity n_{ij} has a binomial distribution with parameters n_j and P_{ij} and so

$$
\begin{aligned}
\mathbf{I}(\boldsymbol{\theta})_{rs} &= -E(\partial^2 l/\partial\theta_s\partial\theta_r) \\
&= -\sum_{j=1}^{J}\sum_{i=1}^{k_j} n_j P_{ij}\frac{P_{ij}\partial^2 P_{ij}/\partial\theta_s\partial\theta_r - (\partial P_{ij}/\partial\theta)(\partial P_{ij}/\partial\theta_s)}{P_{ij}^2} \\
&= -\sum_{j=1}^{J} n_j \sum_{i=1ij}^{k_j} \partial^2 P_{ij}/\partial\theta_s\partial\theta_r - \frac{(\partial P_{ij}/\partial\theta)(\partial P_{ij}/\partial\theta_s)}{P_{ij}}.
\end{aligned}
$$

The first term is the second derivative of 1 and so is zero, leaving the second term.

2.37 The minimum is achieved at $\hat{\theta} = 4.374$. The calculations at the minimum are in Table 1.2.

2.38 (a) Three observations exceed 200. The empirical estimate is $3/20 = 0.15$.
(b) $E(X) = 100\alpha/(\alpha-1) = \bar{x} = 154.5$. $\hat{\alpha} = 154.5/54.5 = 2.835$. $\Pr(X > 200) = (100/200)^{2.835} = 0.140$.

Table 1.2 Calculations for Exercise 2.37

x	$F_n(x)$	$1 - e^{-x/\theta}$	$[F_n(x) - 1 + e^{-x/\theta}]^2$
2	2/9	0.3670	0.0210
5	2/3	0.6812	0.0002
8	1	0.8394	0.0258
		Total	0.0470

(c) $f(x) = \alpha 100^\alpha x^{-\alpha-1}$. $L = \alpha^{20} 100^{20\alpha} (\Pi x_j)^{-\alpha-1}$.

$$l = 20 \log \alpha + 20\alpha \log 100 - (\alpha + 1)\Sigma \log x_j$$
$$l' = 20\alpha^{-1} + 20 \log 100 - \Sigma \log x_j = 0$$
$$\hat{\alpha} = 20/(\Sigma \log x_j - 20 \log 100) = 20/(99.125 - 92.103) = 2.848$$

$\Pr(X > 200) = (100/200)^{2.848} = 0.139$.
(d) The empirical variance is

$$
\begin{aligned}
[1 - F(200)]F(200)/20 &= (100/200)^\alpha [1 - (100/200)^\alpha]/20 \\
&= (0.5^\alpha - 0.25^\alpha)/20.
\end{aligned}
$$

(e) A run of 5,000 bootstrap samples of size 20 produced the results in Table 1.3.

Table 1.3 Calculations for Exercise 2.38

Estimator	Average	Variance
Empirical	0.149	0.00622
Moments	0.133	0.00294
Maximum likelihood	0.138	0.00343

2.39 For minimum modified chi-square estimation the expected number of observations is $G_j(\theta) = n_j[\exp(-c_j/\theta) - \exp(-c_{j-1}/\theta)]$ and for maximum likelihood estimation the two key quantities are $P_j(\theta) = \exp(-c_j/\theta) - \exp(-c_{j-1}/\theta)$ and $P_j'(\theta) = -c_j^{-1} \exp(-c_j/\theta) + c_{j-1}^{-1} \exp(-c_{j-1}/\theta)$. The relevant figures appear in Table 1.4. The minimum modified chi-square estimate is $\hat{\theta} = 89.976$ while the loglikelihood is maximized at $\hat{\theta} = 93.188$. The $P_j'(\theta)$ column totals to zero, indicating that the maximum has been found. The denominator requested is 250 times the sum of the last column, or 0.02701.

Table 1.4 Calculations for Exercise 2.39

Interval	n_j	$G_j(\theta)$	χ^2	$P_j(\theta)$	$n_j \log P_j(\theta)$	$10^3 P_j'(\theta)$	$\dfrac{n_j P_j'(\theta)}{P_j(\theta)}$	$10^6 \dfrac{[P_j'(\theta)]^2}{P_j(\theta)}$
0–25	5	6.84	0.68	0.024	−18.64	−0.962	−.200	38.484
25–50	37	34.51	0.17	0.131	−75.19	−2.140	−.604	34.939
50–75	28	33.98	1.28	0.134	−56.37	−0.747	−.157	4.178
75–100	31	26.34	0.70	0.105	−69.82	−0.089	−.026	0.076
100–125	23	20.04	0.38	0.081	−57.90	0.142	.041	0.251
125–150	9	15.51	4.71	0.063	−24.91	0.214	.031	0.731
150–200	22	22.20	0.00	0.090	−52.91	0.444	.108	2.185
200–250	17	15.01	0.23	0.061	−47.47	0.382	.106	2.386
250–350	15	18.89	1.01	0.077	−38.38	0.566	.110	4.139
350–500	17	15.50	0.13	0.064	−46.81	0.529	.141	4.398
500–750	13	12.91	0.00	0.053	−38.14	0.482	.118	4.374
750–1,000	12	6.75	2.30	0.028	−42.97	0.267	.115	2.549
1,000–1,500	3	6.96	5.22	0.029	−10.65	0.285	.030	2.816
1,500–2,500	5	5.72	0.10	0.024	−18.72	0.241	.051	2.459
2,500–5,000	5	4.38	0.08	0.018	−20.05	0.189	.052	1.972
5,000–10,000	3	2.22	0.20	0.009	−14.07	0.097	.032	1.029
10,000–25,000	3	1.34	0.92	0.006	−15.58	0.059	.032	0.631
25,000–	2	0.90	0.61	0.004	−11.19	0.040	.021	0.427
Total	250	250	18.71	1.000	−659.76	0.000	.000	108.023

1.4   SECTION 2.5

2.40 (a) $f(x) = px^{p-1}$. $\log f(x) = \log p + (p-1)\log x$. $\partial^2 \log f(x)/\partial p^2 = -p^{-2}$. $I(p) = nE(p^{-2}) = np^{-2}$. $Var(\hat{p}) \doteq p^2/n$.
(b) From Exercise 2.29, $\hat{p} = -n/\Sigma \log x_j$. The CI is $\hat{p} \pm 1.96\hat{p}/\sqrt{n}$.
(c) $\mu = p/(1+p)$. $\hat{p}/(1+\hat{p})$. $\partial\mu/\partial p = (1+p)^{-2}$. $Var(\hat{\mu}) \doteq (1+p)^{-4}p^2/n$.
The CI is $\hat{p}(1+\hat{p})^{-1} \pm 1.96\hat{p}(1+\hat{p})^{-2}/\sqrt{n}$.

2.41 (a) $\log f(x) = -\log\theta - x/\theta$. $\partial^2 \log f(x)/\partial\theta^2 = \theta^{-2} - 2\theta^{-3}x$. $I(\theta) = nE(-\theta^{-2} + 2\theta^{-3}X) = n\theta^{-2}$. $Var(\hat{\theta}) \doteq \theta^2/n$.
(b) From Exercise 2.32, $\hat{\theta} = \bar{x}$. The CI is $\bar{x} \pm 1.96\bar{x}/\sqrt{n}$.
(c) $Var(X) = \theta^2$. $\partial Var(X)/\partial\theta = 2\theta$. $\widehat{Var(X)} = \bar{x}^2$. $Var[\widehat{Var(X)}] \doteq (2\theta)^2\theta^2/n = 4\theta^4/n$. The CI is $\bar{x}^2 \pm 1.96(2\bar{x}^2)/\sqrt{n}$.

2.42 $\log f(x) = -(1/2)\log(2\pi\theta) - x^2/(2\theta)$. $\partial^2 \log f(x)/\partial\theta^2 = (2\theta^2)^{-1} - x^2(\theta^3)^{-1}$. $I(\theta) = nE[-(2\theta^2)^{-1} + X^2(\theta^3)^{-1}] = n(2\theta^2)^{-1}$ since $X \sim N(0,\theta)$. Then $MSE(\hat{\theta}) \doteq Var(\hat{\theta}) \doteq 2\theta^2/n \doteq 2\hat{\theta}^2/n = 8/40 = 0.2$.

2.43 (a) $L = F(2)[1-F(2)]^3$. $F(2) = \int_0^2 2\lambda x e^{-\lambda x^2} dx = -e^{-\lambda x^2}\Big|_0^2 = 1 - e^{-4\lambda}$.
$l = \log(1 - e^{-4\lambda}) - 12\lambda$. $\partial l/\partial\lambda = (1 - e^{-4\lambda})^{-1}4e^{-4\lambda} - 12 = 0$. $e^{-4\lambda} = 3/4$.
$\hat{\lambda} = (1/4)\log(4/3)$.
(b) $P_1(\lambda) = 1 - e^{-4\lambda}$, $P_2(\lambda) = e^{-4\lambda}$, $P_1'(\lambda) = 4e^{-4\lambda}$, $P_2'(\lambda) = -4e^{-4\lambda}$.
$I(\lambda) = 4[16e^{-8\lambda}/(1 - e^{-4\lambda}) + 16e^{-8\lambda}/e^{-4\lambda}]$. $\widehat{Var(\hat{\lambda})} = \{4[16(9/16)/(1/4) + 16(9/16)/(3/4)]\}^{-1} = 1/192$.

2.44
$$P_{ij}(\alpha,\theta) = 1 - \left(\frac{\theta}{\theta+c_i}\right)^\alpha - 1 + \left(\frac{\theta}{\theta+c_{i-1}}\right)^\alpha$$
$$= \left(\frac{\theta}{\theta+c_{i-1}}\right)^\alpha - \left(\frac{\theta}{\theta+c_i}\right)^\alpha$$
$$\partial P_{ij}(\alpha,\theta)/\partial\alpha = \left(\frac{\theta}{\theta+c_{i-1}}\right)^\alpha \log\left(\frac{\theta}{\theta+c_{i-1}}\right) - \left(\frac{\theta}{\theta+c_i}\right)^\alpha \log\left(\frac{\theta}{\theta+c_i}\right)$$
$$\partial P_{ij}(\alpha,\theta)/\partial\theta = \frac{\alpha c_{i-1}\theta^{\alpha-1}}{(\theta+c_{i-1})^{\alpha+1}} - \frac{\alpha c_i\theta^{\alpha-1}}{(\theta+c_i)^{\alpha+1}}.$$

The calculations are in Table 1.5 where $P = P_{ij}(\alpha,\theta)$, $P_1 = \partial P_{ij}(\alpha,\theta)/\partial\alpha$, and $P_2 = \partial P_{ij}(\alpha,\theta)/\partial\theta$.

Table 1.5 Calculations for Exercise 2.44

Interval	n_j	n_{ij}	P	P_1	$10^6 P_2$	$\dfrac{n_{ij}P_1}{P}$	$\dfrac{10^3 n_{ij}P_2}{P}$	$\dfrac{n_j P_1^2}{P}$	$\dfrac{n_j P_1 P_2}{P}$	$\dfrac{10^9 n_{ij}P_2^2}{P}$
0–25,000	5,000	2,900	.554	.070	−2.27	366	−11.85	44.15	−.00143	46.30
25,000–50,000	5,000	1,200	.224	−.005	0.30	−27	1.59	0.56	−.00003	1.97
50,000–	5,000	900	.222	−.065	1.97	−263	7.99	94.90	−.00288	87.34
0–25,000	2,000	1,000	.554	.070	−2.27	126	−4.09	17.66	−.00057	18.52
25,000–50,000	2,000	300	.224	−.005	0.30	−7	0.40	0.23	−.00001	0.79
50,000–60,000	2,000	200	.050	−.006	0.23	−24	0.91	1.48	−.00006	2.05
60,000–100,000	2,000	300	.103	−.023	0.76	−67	2.23	10.20	−.00034	11.31
100,000–	2,000	200	.069	−.036	0.98	−104	2.83	37.25	−.00102	27.77
					Total	0	0	206.46	−.00634	196.05

The zeros indicate that the maximum was reached. The information matrix is

$$\begin{bmatrix} 206.4564 & -0.0063398 \\ -0.0063398 & 1.9605 \times 10^{-7} \end{bmatrix}$$

and the estimated covariance matrix is its inverse

$$\begin{bmatrix} 0.692242 & 22{,}385.3 \\ 22{,}385.3 & 728{,}985{,}000 \end{bmatrix}.$$

The mean is $\theta/(\alpha-1)$ and the partial derivatives are $-\theta/(\alpha-1)^2 = -8{,}550.10$, and $1/(\alpha-1) = 0.241066$ when evaluated at the estimators. The variance of the estimate of the mean is

$$\begin{bmatrix} -8{,}550.1 & 0.241066 \end{bmatrix} \begin{bmatrix} 0.692242 & 22{,}385.3 \\ 22{,}385.3 & 728{,}985{,}000 \end{bmatrix} \begin{bmatrix} -8{,}550.1 \\ 0.241066 \end{bmatrix} = 690{,}649.$$

The estimate of the mean is $147{,}130/(5.14825 - 1) = 35{,}468$ and the CI is $35{,}468 \pm 1.96\sqrt{690{,}640} = 35{,}468 \pm 1{,}629.$

2.45 From Exercise 2.30, $l = 81.61837(\alpha - 1) - 38{,}000\theta^{-1} - 10\log\Gamma(\alpha) - 10\alpha\log\theta$. Also, $\hat{\alpha} = 6.341$ and $\hat{\theta} = 599.3$. Using $v = 4$, we have

$$\frac{\partial^2 l(\alpha,\theta)}{\partial\alpha^2} \doteq \frac{l(6.3416341,599.3) - 2l(6.341,599.3) + l(6.3403659,599.3)}{(.0006341)^2} = -1.70790$$

$$\frac{\partial^2 l(\alpha,\theta)}{\partial\alpha\partial\theta} \doteq \frac{l(6.34131705,599.329965) - l(6.34131705,599.270035)}{-l(6.34068295,599.329965) + l(6.34068295,599.270035)}{(.0006341)(.05993)}$$

$$= 0.0166861$$

$$\frac{\partial^2 l(\alpha,\theta)}{\partial\theta^2} \doteq \frac{l(6.341,599.35993) - 2l(6.341,599.3) + l(6.341,599.25007)}{(.05993)^2} = -0.000176536$$

$$I(\hat{\alpha},\hat{\theta}) = \begin{bmatrix} 1.70790 & -0.0166861 \\ -0.0166861 & 0.000176536 \end{bmatrix}$$

and its inverse is

$$V\hat{a}r = \begin{bmatrix} 7.64976 & -723.055 \\ -723.055 & 74{,}007.7 \end{bmatrix}.$$

The mean is $\alpha\theta$ and so the derivative vector is $\begin{bmatrix} 599.3 & 6.341 \end{bmatrix}$. The variance of $\widehat{\alpha\theta}$ is estimated as $227{,}763$ and a 95% CI is $3{,}800 \pm 1.97\sqrt{227{,}763} = 3{,}800 \pm 935$. For the next observation, the variance is $\hat{\alpha}\hat{\theta}^2 + 227{,}763 = 2{,}505{,}200$ and the PI is $3{,}800 \pm 3{,}102$.

2.46 From Exercise 2.31, $\hat{\mu} = 7.33429$ and $\hat{\sigma} = 0.753263$. From Example 2.24, the covariance matrix is $\begin{bmatrix} \sigma^2/5 & 0 \\ 0 & \sigma^2/10 \end{bmatrix}$ and inserting the parameter estimates gives $\begin{bmatrix} 0.113481 & 0 \\ 0 & 0.0567405 \end{bmatrix}$. $h(\mu,\sigma) = \Pr(X > 4{,}500) =$

$1 - \Phi\left(\frac{\log 4,500 - \mu}{c}\right)$. The derivatives are $\partial h/\partial\mu = \sigma^{-1}\phi\left(\frac{\log 4,500 - \mu}{\sigma}\right)$ and $\partial h/\partial\sigma = \frac{\log 4,500 - \mu}{\sigma^2}\phi\left(\frac{\log 4,500 - \mu}{\sigma}\right)$ where $\phi(x)$ is the standard normal *pdf*. Inserting the parameter estimates in the derivatives gives $\begin{bmatrix} 0.190400 & 0.272331 \end{bmatrix}$ and so the variance of the probability estimate is 0.00832204 and the CI is $0.076 \pm 1.96\sqrt{0.00832204} = 0.076 \pm 0.179$.

2.47 The method of moments estimator is $\bar{X}\sqrt{2/\pi}$ and its variance is

$$2Var(X)/(n\pi) = 2(2\beta^2 - \pi\beta^2/2)/(n\pi) = \beta^2(4/\pi - 1)/n.$$

Inserting the *mle* of $\hat{\beta} = 3.20031$ gives a variance of 0.559703. The estimate of the mean is $\hat{\beta}\sqrt{\pi/2}$ and inserting the method of moments estimate gives 4.2. Then

$$Var(\hat{\beta}\sqrt{\pi/2}) = 0.559703(\pi/2) = 0.879179$$

and the 95% CI is 4.2 ± 1.84.

For the *mle*, $\log f(x) = -2\log\beta + \log x - x^2/(2\beta)^2$. The second derivative with respect to β is $2\beta^{-2} - 3x^2\beta^{-4}$ and the negative expected value is $I(\beta) = n4\beta^{-2}$ and the variance is $\beta^2/(4n)$ which is 0.512099 when $\hat{\beta}$ is inserted. The estimate of the mean is 4.01 and its variance is $0.512099(\pi/2) = 0.804403$ and the CI is 4.01 ± 1.76.

2.48 $\hat{\alpha} = 3.8629$. $\log f(x) = \log\alpha + \alpha\log\lambda - (\alpha+1)\log(\lambda+x)$. $\partial^2\log f(x)/\partial\alpha^2 = -\alpha^{-2}$. $I(\alpha) = n\alpha^{-2}$. $Var(\hat{\alpha}) \doteq \alpha^2/n$. Inserting the estimate gives 2.9844.

$$
\begin{aligned}
E(X \wedge 500) &= \int_0^{500} x\alpha 1,000^\alpha(1,000 + x)^{-\alpha-1}dx \\
&\quad + 500\int_{500}^\infty \alpha 1,000^\alpha(1,000 + x)^{-\alpha-1}dx \\
&= \frac{1,000}{\alpha - 1} - (2/3)^\alpha\frac{1,500}{\alpha - 1}.
\end{aligned}
$$

Evaluated at $\hat{\alpha}$ it is 239.88. The derivative with respect to α is

$$-\frac{1,000}{(\alpha - 1)^2} + (2/3)^\alpha\frac{1,500}{(\alpha - 1)^2} - (2/3)^\alpha\frac{1,500}{\alpha - 1}\log(2/3)$$

which is -39.428 when evaluated at $\hat{\alpha}$. The variance of the LEV estimator is $(-39.4298)^2(2.9844) = 5,639.45$ and the CI is 239.88 ± 133.50.

2.49 $f(x) = \alpha 100^\alpha x^{-\alpha-1}$.

$$
\begin{aligned}
\log f(x) &= \log\alpha + \alpha\log 100 - (\alpha + 1)\log x \\
\partial^2\log f(x)/\partial\alpha^2 &= -\alpha^{-2} \\
Var(\hat{\alpha}) &\doteq \alpha^2/n \doteq 2.848^2/20 = 0.40556
\end{aligned}
$$

$h(\alpha) = \Pr(X > 200) = (0.5)^\alpha$. $h'(\alpha) = (0.5)^\alpha \log(0.5) \doteq -0.096270$. The variance of the *mle* of $\Pr(X > 200)$ is approximately $(-0.09627)^2(0.40556) = 0.00376$ which is similar to the bootstrap estimate of 0.00434 from Exercise 2.38.

2.50 From Exercise 2.39, $Var(\hat{\theta}) \doteq 0.02701$ and $\hat{\theta} = 93.188$. For

$$\begin{aligned}
h(\theta) &= \Pr(X > 10{,}000) = 1 - \exp(-\theta/10{,}000), \\
h(\hat{\theta}) &= 0.0092755, \\
h'(\theta) &= (10{,}000)^{-1}\exp(-\theta/10{,}000), \text{ and} \\
h'(\hat{\theta}) &= 0.000099072.
\end{aligned}$$

The variance of the *mle* of $\Pr(X > 10{,}000)$ is approximately

$$(0.000099072)^2(0.0092755) = 2.6511 \times 10^{-10}.$$

The CI is 0.00928 ± 0.00003.

2.51 From Example 2.24, the diagonal elements of the information matrix are n/σ^2 and $2n/\sigma^2$ while the off diagonal element is zero. Substituting parameter estimates in the second derivatives of the negative loglikelihood gives:

$-\partial^2 l/\partial\mu^2 = n/\sigma^2$ which matches the true information value even before the estimates are inserted.

$-\partial^2 l/\partial\theta\partial\mu = 2\Sigma(\log x_j - \mu)/\sigma^3$. When the *mle* $\hat{\mu} = n^{-1}\Sigma\log x_j$ is inserted, the result is zero.

$-\partial^2 l/\partial\sigma^2 = -n/\sigma^2 + 3\Sigma(\log x_j - \mu)^2/\sigma^4$ and for the *mle*, $\Sigma(\log x_j - \hat{\mu})^2 = n\hat{\sigma}^2$ and so inserting the *mle* gives $-n/\hat{\sigma}^2 + 3n\hat{\sigma}^2/\hat{\sigma}^4 = 2n/\hat{\sigma}^2$.

2.52 It turns out that for the Pareto distribution, the two matrices agree except for the lower right corner. The second derivative of the negative loglikelihood is $-\partial^2 l/\partial\theta^2 = n\alpha\theta^{-2} - (\alpha + 1)\Sigma(x_j + \theta)^{-2}$. The corresponding member of the information matrix is

$$\begin{aligned}
n\alpha\theta^{-2} - (\alpha + 1)nE[(X + \theta)^{-2}] &= n\alpha\theta^{-2} \\
&\quad -n(\alpha + 1)\int_0^\infty \alpha\theta^\alpha(x + \theta)^{-\alpha - 3}dx \\
&= n\alpha\theta^{-2} - n(\alpha + 1)\alpha\theta^{-2}/(\alpha + 2) \\
&= n\theta^{-1}\alpha/(\alpha + 2).
\end{aligned}$$

The two expressions need not match when the *mle*'s are inserted as the information is a function of $\Sigma(x_j + \theta)^{-1}$ but not $\Sigma(x_j + \theta)^{-2}$.

1.5 SECTION 2.6

2.53 (a) With 10% inflation the boundaries increase by 10%. For the first eleven groups the contribution to the mean is $\sum_{i=1}^{11} 1.1(n_i/217)(c_{i-1}+c_i)/2 = 26{,}993.088$. For the twelfth group we have

$$\int_{247,500}^{300,000} x\frac{4}{217(82,500)}\,dx + \int_{300,000}^{330,000} 300{,}000\frac{4}{217(82,500)}\,dx = 5{,}222.04.$$

For the last group the contribution is $300{,}000(3/217) = 4{,}147.465$ for a total of $36{,}362.59$. To evaluate the impact of this increase, the mean prior to inflation is

$$\sum_{j=1}^{12}(n_j/217)(c_{j-1}+c_j)/2 + 300{,}000(3/217) = 33{,}525.35$$

for a percentage increase of 8.46%.
(b) With a deductible of 1,000 the mean is

$$\int_{1,000}^{2,500}(x-1{,}000)\frac{41}{217(2,500)}\,dx + \sum_{j=2}^{10}\frac{n_j}{217}\left(\frac{c_{j-1}+c_j}{2}-1{,}000\right)$$
$$+299{,}000(3/217) = 32{,}563.13$$

for a percentage decrease of 2.87%.

2.54 From Example 2.36 the data set with a limit of 300,000 had a loglike-lihood of -498.29 and the set with a limit of 500,000 had a loglikelihood of $-1{,}367.12$. The two new sets had loglikelihoods of -222.16 (100,000 limit, parameter estimates are $\hat{\mu} = 9.32126$ and $\hat{\sigma} = 1.70973$) and -821.33 (1,000,000, $\hat{\mu} = 9.48117$, $\hat{\sigma} = 1.71624$) limit. When all four data sets are combined the loglikelihood is $-2{,}919.26$ ($\hat{\mu} = 9.39305$, $\hat{\sigma} = 1.67179$). The sum of the four individual loglikelihoods is $-2{,}917.90$. Twice the difference is 2.72. At a 5% significance level, the chi-square critical value with 6 degrees of freedom is 12.59, indicating that the hypothesis that all four samples could be from the same lognormal population is accepted. Also, the p-value is 0.8407.

2.55

$$F_Y(y) = \Phi\left[\frac{y/c-\mu}{\mu}\left(\frac{\theta c}{y}\right)^{1/2}\right] + \exp\left(\frac{2\theta}{\mu}\right)\Phi\left[-\frac{y/c+\mu}{\mu}\left(\frac{\theta c}{y}\right)^{1/2}\right]$$
$$= \Phi\left[\frac{y-c\mu}{c\mu}\left(\frac{\theta c}{y}\right)^{1/2}\right] + \exp\left(\frac{2c\theta}{c\mu}\right)\Phi\left[-\frac{y+c\mu}{c\mu}\left(\frac{\theta c}{y}\right)^{1/2}\right]$$

and so Y is inverse Gaussian with parameters $c\mu$ and $c\theta$. Because it is still inverse Gaussian, it is a scale family. Because both μ and θ change there is no scale parameter.

2.56 $F_Y(y) = \Gamma[\alpha; (y/c)/\theta] = \Gamma[\alpha; y/(c\theta)]$ and so Y is gamma with parameters α and $c\theta$.

2.57
$$
\begin{aligned}
E(Y^2) &= \int_d^u \alpha^2 (x-d)^2 f(x)dx + \int_u^\infty \alpha^2 (u-d)^2 f(x)dx \\
&= \alpha^2 \{ E[(X \wedge u)^2] - u^2[1 - F(u)] - E[(X \wedge d)^2] \\
&\quad + d^2[1 - F(d)] - 2dE(X \wedge u) + 2du[1 - F(u)] \\
&\quad + 2dE(X \wedge d) - 2d^2[1 - F(d)] + d^2[F(u) - F(d)] \\
&\quad + (u^2 - 2ud + d^2)[1 - F(u)] \} \\
&= \alpha^2 \{ E[(X \wedge u)^2] - E[(X \wedge d)^2] \\
&\quad - 2dE(x \wedge u) + 2dE(X \wedge d) \}
\end{aligned}
$$

2.58 The empirical mean residual life is

$$
e_{\hat{X}}(d) = \frac{\left[\sum_{x_j > d} (x_j - d) \right]}{\#losses > d}.
$$

The results are

$$
e_{\hat{X}}(1,000) = [990,000 - 1,000(180)]/180 = 4,500
$$

and $e_{\hat{X}}(3,000) = 4,475$, $e_{\hat{X}}(5,000) = 4,507$, $e_{\hat{X}}(7,000) = 4,520$, $e_{\hat{X}}(9,000) = 4,500$. $e_{\hat{X}}(d)$ is essentially constant. From Example 2.31 the Pareto distribution has a straight line with positive slope, so the data is not consistent. (Note - the exponential distribution has a constant mean residual life)

2.59
$$
\begin{aligned}
LER(d) &= E(X \wedge d)/E(X) = 1 - \left(\frac{\theta}{\theta + d} \right)^{\alpha - 1} \\
LER(500) &= 1 - \left(\frac{2,000}{2,000 + 500} \right)^{2-1} = 0.2.
\end{aligned}
$$

2.60 1993 expected payment/loss is

$$
(0 + 500 + 1,500 + 2,500 + 3,500 + 4,500)/6 = 2,083\frac{1}{3}.
$$

1994 expected payment/loss is

$$
(0 + 600 + 1,650 + 2,700 + 3,750 + 4,800)/6 = 2,250.
$$

The percentage increase is $(2,250/2,083\frac{1}{3} - 1)100 = 8\%$.

2.61 For 1994, $\mu = 10 + \log(1.1) = 10.0953$ and $\sigma = \sqrt{5}$.

$$
\begin{aligned}
E(X \wedge 2{,}000{,}000) &= e^{10.0953+5/2}\Phi\left(\frac{\log 2{,}000{,}000 - 10.0953 - 5}{\sqrt{5}}\right) \\
&\quad +2{,}000{,}000\left[1 - \Phi\left(\frac{\log 2{,}000{,}000 - 10.0953}{\sqrt{5}}\right)\right] \\
&= 165{,}456
\end{aligned}
$$

2.62 $e_{\hat{X}}(250) = (300 - 250)/1 = 50.$

2.63 $Y = cX$.

$$
\begin{aligned}
F_Y(y) &= \Pr(Y \le y) = \Pr(X \le y/c) \\
&= \Phi\left(\frac{y/c - \mu}{\sigma}\right) = \Phi\left(\frac{y - c\mu}{c\sigma}\right).
\end{aligned}
$$

Therefore, Y is normal with mean $c\mu$ and variance $c^2\sigma^2$, which are 1,050 and 11,025.

2.64 $Y = 1.03X$. $F_X(x) = x/2{,}500$. $F_Y(y) = y/2{,}575$.

$$
\begin{aligned}
E(Y) &= \int_0^{2{,}575} \frac{y}{2{,}575}\,dy = 1{,}287.5 \\
E(Y \wedge 100) &= \int_0^{100} \frac{y}{2{,}575}\,dy + 100\frac{2{,}475}{2{,}575} = 98.058.
\end{aligned}
$$

$LER = \frac{98.058}{1{,}287.5} = 0.076.$

2.65 From Example 2.31, $e_X(d) = (d + \theta)/(\alpha - 1)$ and so $e_X(2\theta)/e_X(\theta) = (2\theta + \theta)/(\theta + \theta) = 1.5.$

2.66 From Exercise 2.38, $\hat{\alpha} = 2.848$. Also, $\theta = 100$.
(a) $E(X) = \alpha\theta/(\alpha - 1) = 154.11.$
(b) $E(X \wedge 200) = \alpha\theta/(\alpha - 1) - \theta^\alpha/[(\alpha - 1)200^{\alpha-1}] = 139.08.$ $LER = 0.9025.$
(c)
$$
\begin{aligned}
E(\text{payment/loss}) &= 0.8E[E(X) - E(X \wedge 500)] \\
&= 0.8\theta^\alpha/[(\alpha - 1)500^{\alpha-1}] = 2.2115.
\end{aligned}
$$

$E(\text{payment/payment}) = 2.2115/\Pr(X > 500) = 2.2115/(\theta/500)^\alpha = 216.45.$

From the data, there were no losses above 500, so the empirical average payment/loss is zero and the average payment/payment is undefined.

(d)
$$F_Y(y) = \Pr(X \le y/1.05) = 1 - \left(\frac{100}{y/1.05}\right)^\alpha$$
$$= 1 - \left(\frac{105}{y}\right)^\alpha.$$

So Y is Pareto with $\theta = 105$ and α unchanged. $E(Y) = 2.848(105)/1.848 = 161.82$.

(e) Using the same formula as in (b), $E(Y \wedge 200) = 144.55$ and $LER = 0.8933$.

2.67 $\hat{\alpha} = 0.432947$ and $\hat{\theta} = 1{,}705.39$. The per loss expected value is $0.8[E(X \wedge 25{,}000) - E(X \wedge 5{,}000)]$. From Appendix A,

$$E(X \wedge 25{,}000) = 0.432947(1705.39)\Gamma(1.432947; 25{,}000/1705.39)]$$
$$+25{,}000[1 - \Gamma(0.432947; 25{,}000/1{,}705.39)]$$
$$= 738.34.$$

Similarly, $E(X \wedge 5{,}000) = 719.81$ and the answer is $.8(738.34 - 719.81) = 14.82$.

2.68 (a) For the 10,000 limit, $\hat{\theta} = 1{,}358$, $\hat{\tau} = 0.5272$, $E(X \wedge 20{,}000) = 1{,}982$, and $NLL = 2{,}311.13$ (negative loglikelihood). For the 25,000 limit, $\hat{\theta} = 1{,}584$, $\hat{\tau} = 0.5067$, $E(X \wedge 25{,}000) = 2{,}829$, $NLL = 4{,}278.05$. The increased limits factor is $2{,}829/1{,}982 = 1.427$.

(b) When the data is put together in one sample, $\hat{\theta} = 1{,}894$, $\hat{\tau} = 0.4807$, and $NLL = 54{,}260.96$. The increased limits factor is $3{,}459/2{,}570 = 1.346$.

(c) $\hat{\tau} = 0.4892$ while the five scale parameters are 1,168, 1,293, 1,540, 1,660, and 2,275 and $NLL = 54{,}079.38$. The increased limits factor is $2{,}877/2{,}007 = 1.433$.

(d) For the other three limits we have — 5,000: $\hat{\theta} = 1{,}227$, $\hat{\tau} = 0.5521$, $NLL = 6{,}537.94$; 50,000: $\hat{\theta} = 1{,}633$, $\hat{\tau} = 0.4810$, $NLL = 6{,}254.07$; 100,000: $\hat{\theta} = 2{,}231$, $\hat{\tau} = 0.4794$, $NLL = 34{,}666.32$. The total of the five NLL's is 54,047.52. For model (b) versus model (c) the test statistic is $2(54{,}260.96 - 54{,}079.38) = 363.16$ There are $6 - 2 = 4$ degrees of freedom and the null hypothesis of model (b) is rejected. For model (a) versus model (c) the test statistic is $2(54{,}079.38 - 54{,}047.52) = 63.72$ with $10 - 6 = 4$ degrees of freedom and the null hypothesis of model (c) is rejected. Conclusion – choose the most complicated model where each group has its own model.

1.6 SECTION 2.7

2.69 Means:

Gamma : $0.2(500) = 100$

$$\text{Lognormal} \quad : \quad \exp(3.70929 + 1.33856^2/2) = 99.9999$$
$$\text{Pareto} \quad : \quad 150/(2.5 - 1) = 100.$$

Second moments:

$$\text{Gamma} \quad : \quad 500^2(0.2)(1.2) = 60,000$$
$$\text{Lognormal} \quad : \quad \exp[2(3.70929) + 2(1.33856)^2] = 59,999.88$$
$$\text{Pareto} \quad : \quad 150^2(2)/[1.5(0.5)] = 60,000.$$

The density functions are:

$$\text{Gamma} \quad : \quad 0.754921x^{-0.8}e^{-0.002x}$$

$$\text{Lognormal} \quad : \quad (2\pi)^{-1/2}(1.338566x)^{-1}\exp\left[-\frac{1}{2}\left(\frac{\log x - 3.70929}{1.338566}\right)^2\right]$$

$$\text{Pareto} \quad : \quad 688,919(x+150)^{-3.5}.$$

The gamma and lognormal densities are equal when $x = 4,341$ while the lognormal and Pareto densities are equal when 9,678. Numerical evaluation indicates that the ordering is as expected.

2.70 $F_Y(y) = 1 - (1 + y/\theta)^{-\alpha} = 1 - \left(\frac{\theta}{\theta+y}\right)^\alpha$. This is the *cdf* of the Pareto distribution. $f_Y(y) = dF_Y(y)/dy = \frac{\alpha\theta^\alpha}{(\theta+y)^{\alpha+1}}$.

2.71 Inverse: $F_Y(y) = 1 - \left[1 - \left(\frac{\theta}{\theta+y^{-1}}\right)^\alpha\right] = \left(\frac{y}{y+\theta^{-1}}\right)^\alpha$. From Appendix A this is the inverse Pareto distribution with $\tau = \alpha$ and $\theta = 1/\theta$. Transformed: $F_Y(y) = 1 - \left(\frac{\theta}{\theta+y^\tau}\right)^\alpha$. This is the Burr distribution with $\alpha = \alpha$, $\gamma = \tau$, and $\theta = \theta^{1/\gamma}$. Inverse transformed: $F_Y(y) = 1 - \left[1 - \left(\frac{\theta}{\theta+y^{-\tau}}\right)^\alpha\right] = \left[\frac{y^\tau}{y^\tau+(\theta^{-1/\tau})^\tau}\right]$. This is the inverse Burr distribution with $\tau = \alpha$, $\gamma = \tau$, and $\theta = \theta^{-1/\tau}$.

2.72 $F_Y(y) = 1 - \frac{(y^{-1}/\theta)^\gamma}{1+(y^{-1}/\theta)^\gamma} = \frac{1}{1+(y^{-1}/\theta)^\gamma} = \frac{(y\theta)^\gamma}{1+(y\theta)^\gamma}$. This is the loglogistic distribution with γ unchanged and $\theta = 1/\theta$.

2.73 $F_Y(y) = \Phi\left[\frac{\log(y/\theta)-\mu}{\sigma}\right] = \Phi\left[\frac{\log y - \log\theta - \mu}{\sigma}\right]$ which is the *cdf* of a lognormal distribution with $\mu = \ln\theta + \mu$, and $\sigma = \sigma$.

2.74 $f_Y(y) = \exp(-|(\log y)/\theta|)/2\theta y$, For $x \le 0$, $F_X(x) = \frac{1}{2\theta}\int_{-\infty}^x e^{t/\theta}dt = \frac{1}{2}e^{t/\theta}\Big|_{-\infty}^x = \frac{1}{2}e^{x/\theta}$. For $x > 0$ it is $\frac{1}{2} + \frac{1}{2\theta}\int_0^x e^{-t/\theta}dt = 1 - \frac{1}{2}e^{-x/\theta}$. With exponentiation the two descriptions are $F_Y(y) = \frac{1}{2}e^{\ln y/\theta}$, $0 < y \le 1$, and $F_Y(y) = 1 - \frac{1}{2}e^{-\ln y/\theta}$, $y \ge 1$.

2.75 $X|\Theta = \theta$ has *pdf*

$$f_{X|\Theta}(x|\theta) = \frac{\tau \left[(x/\theta)^{\tau}\right]^{\alpha} \exp[-(x/\theta)^{\tau}]}{x\Gamma(\alpha)}$$

and Θ has *pdf*

$$f_{\Theta}(\theta) = \frac{\tau \left[(\delta/\theta)^{\tau}\right]^{\beta} \exp[-(\delta/\theta)^{\tau}]}{\theta\Gamma(\beta)}.$$

The mixed distribution has *pdf*

$$
\begin{aligned}
f(x) &= \int_{0}^{\infty} \frac{\tau \left[(x/\theta)^{\tau}\right]^{\alpha} \exp[-(x/\theta)^{\tau}]}{x\Gamma(\alpha)} \frac{\tau \left[(\delta/\theta)^{\tau}\right]^{\beta} \exp[-(\delta/\theta)^{\tau}]}{\theta\Gamma(\beta)} d\theta \\
&= \frac{\tau^2 x^{\tau\alpha} \delta^{\tau\beta}}{x\Gamma(\alpha)\Gamma(\beta)} \int_{0}^{\infty} \theta^{-\tau\alpha-\tau\beta-1} \exp[-\theta^{-\tau}(x^{\tau} + \delta^{\tau})] d\theta \\
&= \frac{\tau^2 x^{\tau\alpha} \delta^{\tau\beta}}{x\Gamma(\alpha)\Gamma(\beta)} \frac{\Gamma(\alpha+\beta)}{\tau(x^{\tau} + \delta^{\tau})^{\alpha+\beta}} = \frac{\Gamma(\alpha+\beta)\tau x^{\tau\alpha-1}\delta^{\tau\beta}}{\Gamma(\alpha)\Gamma(\beta)(x^{\tau} + \delta^{\tau})^{\alpha+\beta}}
\end{aligned}
$$

which is a transformed beta *pdf* with (using the parameterization in Appendix A) $\gamma = \tau$, $\tau = \alpha$, $\alpha = \beta$, and $\theta = \delta$.

2.76 The generalized Pareto distribution is the transformed beta distribution with $\gamma = 1$. The limiting distribution is then transformed gamma with $\tau = 1$, which is a gamma distribution. The gamma parameters are $\alpha = \tau$ and $\theta = \xi$.

2.77 The maximum likelihood estimate of the exponential parameter is $\hat{\theta} = 32{,}874.8$ and the loglikelihood is -548.722. For the Weibull distribution the parameter estimates are $\hat{\tau} = 0.616171$ and $\hat{\theta} = 22{,}753.6$ and the loglikelihood is -501.626. Twice the difference is 94.192. When compared with the chi-square (with one degree of freedom) critical value at a 1% significance level of 6.635, the null hypothesis favoring the exponential model is rejected in favor of the Weibull model.

2.78 To be a legitimate *cdf* the function must be non-decreasing, right-continuous, and span the values from zero to one. Because the a_i's sum to one, the limit as y goes to infinity will be one and similarly the limit as y goes to negative infinity will be zero. A linear combination of right-continuous functions must also be right-continuous. Because the a_i's are all positive, the function must be non-decreasing.

2.79 For a single exponential distribution the estimate is $\hat{\theta} = 32{,}874.8$ and $NLL = 548.722$. For the mixture with *cdf* $F(x) = 1 - pe^{-x/\theta_1} - (1-p)e^{-x/\theta_2}$ the *mle's* are $\hat{\theta}_1 = 78{,}615.8$, $\hat{\theta}_2 = 9{,}209.40$, and $\hat{p} = 0.35651$ with $NLL =$

500.537. The likelihood ratio test statistic is $2(548.722 - 500.537) = 96.37$. With two degrees of freedom, the p-value is very small. The mixture distribution is preferred.

2.80 From Exercise 2.39, $\hat{\theta} = 93.188$ and $NLL = 659.76$. From Exercise 2.67, $\hat{\theta} = 94.35$, $\hat{\tau} = 0.9639$, and $NLL = 659.51$. The likelihood ratio test statistic is $2(659.76 - 659.51) = 0.5$. With one degree of freedom, the p-value is 0.4795 and the null hypothesis (exponential model) is selected. There is no evidence to support using the additional parameter.

2.81 $F(x) = \int_1^x 3t^{-4}dt = 1 - x^{-3}$. $Y = 1.1X$. $F_Y(y) = 1 - (x/1.1)^{-3}$. $\Pr(Y > 2.2) = 1 - F_Y(2.2) = (2.2/1.1)^3 = 0.125$.

2.82 (a) $F(x) = 1 - \exp\left(-\int_0^x Bc^t dt\right) = 1 - \exp\left(-\frac{Bc^x}{\log c} + \frac{B}{\log c}\right)$. $f(x) = Bc^x \exp\left(-\frac{Bc^x}{\log c} + \frac{B}{\log c}\right)$.

(b) Let $Y = \theta X$. Then $F_Y(y) = 1 - \exp\left(-\frac{Bc^{x/\theta}}{\log c} + \frac{B}{\log c}\right)$. Let $c^* = c^{1/\theta}$ and $B^* = B/\theta$. Then $F_Y(y) = 1 - \exp\left(-\frac{B^*c^{*x}}{\log c^*} + \frac{B^*}{\log c^*}\right)$ and so it is a scale family.

(c) For the Gompertz, $\lambda(x) = Bc^x$ which is an increasing function of x and so the distribution is IFR. For the Weibull, $\lambda(x) = \theta\tau x^{\tau-1}$ which is also IFR. The gamma distribution is IFR when $\alpha > 1$ and so all are equal except for the gamma with $\alpha < 1$ which is DFR and has a heavier tail.

(d) The maximum likelihood estimates are $\hat{B} = 0.01046$ and $\hat{c} = 1.017$. The histogram and *pdf* appear in Figure 1.3.

2.83 Hold α constant and let $\theta\tau^{1/\gamma} \to \xi$. Then let $\theta = \xi\tau^{-1/\gamma}$. Then

$$
\begin{aligned}
f(x) &= \frac{\Gamma(\alpha+\tau)\gamma x^{\gamma\tau-1}}{\Gamma(\alpha)\Gamma(\tau)\theta^{\gamma\tau}(1+x^\gamma\theta^{-\gamma})^{\alpha+\tau}} \\
&= \frac{e^{-\alpha-\tau}(\alpha+\tau)^{\alpha+\tau-1}(2\pi)^{1/2}\gamma x^{\gamma\tau-1}}{\Gamma(\alpha)e^{-\tau}\tau^{\tau-1}(2\pi)^{1/2}\xi^{\gamma\tau}\tau^{-\tau}(1+x^\gamma\xi^{-\gamma}\tau)^{\alpha+\tau}} \\
&= \frac{e^{-\alpha}\left(1+\frac{\alpha}{\tau}\right)^{\alpha+\tau-1}\gamma x^{-\gamma\alpha-1}}{\Gamma(\alpha)\tau^{-\alpha-\tau}\xi^{\gamma(\tau+\alpha)}\xi^{-\gamma\alpha}x^{-\gamma(\tau+\alpha)}(1+x^\gamma\xi^{-\gamma}\tau)^{\alpha+\tau}} \\
&= \frac{e^{-\alpha}\left(1+\frac{\alpha}{\tau}\right)^{\alpha+\tau-1}\gamma x^{-\gamma\alpha-1}}{\Gamma(\alpha)\xi^{-\gamma\alpha}\left[1+\frac{(\xi/x)^\gamma}{\tau}\right]^{\alpha+\tau}} \\
&\to \frac{\gamma\xi^{\gamma\alpha}}{\Gamma(\alpha)x^{\gamma\alpha+1}e^{(\xi/x)^\gamma}}.
\end{aligned}
$$

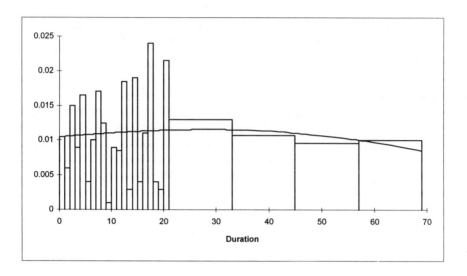

Fig. 1.3 Histogram and *pdf* for Exercise 2.82

2.84 $\gamma = 1$, $\tau = 3$ and $\theta = 50(\alpha + 1)$ for the mode to be 100. Then $\xi = \lim_{\alpha \to \infty} \theta \alpha^{-1/\gamma} = \lim_{\alpha \to \infty} 50(\alpha + 1)\alpha^{-1} = 50$. The limiting distribution is gamma with $\alpha = 3$ and $\theta = 50$.

2.85 $\gamma = 1$, $\alpha = 3$ and $\theta = 200/\tau$ for the mean to be 100. Then, from Exercise 2.83, $\xi = \lim_{\tau \to \infty} \theta \tau^{1/\gamma} = \lim_{\tau \to \infty} 200\tau^{-1}\tau = 200$. The limiting distribution is inverse gamma with $\alpha = 3$ and $\theta = 200$.

2.86 From (2.17)

$$
\begin{aligned}
e(x) &= \int_x^\infty \frac{S(t)}{S(x)} dt \\
&= \int_x^\infty \frac{\exp[-\int_0^t \lambda(y)dy]}{\exp[-\int_0^x \lambda(y)dy]} dt \\
&= \int_x^\infty \exp\left[-\int_x^t \lambda(y)dy\right] dt.
\end{aligned}
$$

If $\lambda(x) \geq \lambda$, then $\int_x^t \lambda(y)dy \geq \int_x^t \lambda dy = \lambda(t - x)$, and then

$$
e(x) \leq \int_x^\infty \exp[-\lambda(t - x)]dt = 1/\lambda.
$$

As well, $S(x) = \exp[-\int_0^x \lambda(y)dy] \leq \exp(-\int_0^x \lambda dy) = \exp(-\lambda x)$. Next, note that from (2.20) $\lambda_e(x) = [e(x)]^{-1} \geq \lambda$ and so $S_e(x) = \exp[-\int_0^x \lambda_e(x)dx] \leq \exp(-\int_0^x \lambda dx) = \exp(-\lambda x)$. Similar arguments provide the other inequalities.

2.87 (a) We have

$$xS(x) + E(X)S_e(x) = xS(x) + \int_x^\infty S(t)dt$$

Use integration by parts to obtain

$$\int_x^\infty S(t)dt = tS(t)|_x^\infty - \int_x^\infty tS'(t)dt = -xS(x) + \int_x^\infty ydF(y)$$

and so

$$xS(x) + E(X)S_e(x) = \int_x^\infty ydF(y).$$

Then, from (2.21)

$$\int_x^\infty ydF(y) = xS(x) + E(X)e(x)S(x)/e(0) = xS(x) + e(x)S(x)$$

since $e(0) = E(X)$. Then

$$S(x) = \frac{\int_x^\infty ydF(y)}{x + e(x)}.$$

Because $E(X) = \int_0^\infty ydF(y) \geq \int_x^\infty ydF(y)$, we have

$$S(x) \leq \frac{E(X)}{x + e(x)}.$$

From the second equation in this Exercise

$$[x + e(x)]S(x) = \int_x^\infty ydF(y).$$

Multiply both sides by $S_e(x)E(X)/S(x)$ to obtain

$$[x + e(x)]S_e(x)E(X) = \frac{S_e(x)E(X)}{S(x)} \int_x^\infty ydF(y).$$

Then use (2.21) to replace $S_e(x)E(X)/S(x)$ with $e(x)$ to obtain

$$S_e(x) = \frac{e(x)\int_x^\infty ydF(y)}{E(X)[x + e(x)]}.$$

Obtain the final inequality by noting that $E(X) \geq \int_x^\infty ydF(y)$ and so

$$S_e(x) = \frac{e(x)}{x + e(x)}.$$

2.88 $DFR \Rightarrow S(x)$ is log convex which implies that $S(x+y)/S(x)$ is increasing in x for fixed y. Therefore,

$$\frac{S(0+y)}{S(0)} \leq \frac{S(x+y)}{S(x)}.$$

Since $S(0) = 1$, we have $S(y)S(x) \leq S(x+y)$ which implies NWU.

Similarly, $IFR \Rightarrow S(x)$ is log concave which implies that $S(x+y)/S(x)$ is decreasing in x for fixed y. Therefore,

$$\frac{S(0+y)}{S(0)} \geq \frac{S(x+y)}{S(x)}.$$

Since $S(0) = 1$, we have $S(y)S(x) \geq S(x+y)$ which implies NBU.

2.89 (a) $NWU \Rightarrow S(x)S(y) \leq S(x+y)$. Under NWU,

$$
\begin{aligned}
S(x)S_e(y) &= \frac{S(x)\int_y^\infty S(t)dt}{e(0)} = \frac{\int_y^\infty S(x)S(t)dt}{e(0)} \\
&\leq \frac{\int_y^\infty S(x+t)dt}{e(0)} = \frac{\int_{x+y}^\infty S(t)dt}{e(0)} = S_e(x+y)
\end{aligned}
$$

which implies $NWUC$. Similarly, under NBU,

$$
\begin{aligned}
S(x)S_e(y) &= \frac{S(x)\int_y^\infty S(t)dt}{e(0)} = \frac{\int_y^\infty S(x)S(t)dt}{e(0)} \\
&\geq \frac{\int_y^\infty S(x+t)dt}{e(0)} = \frac{\int_{x+y}^\infty S(t)dt}{e(0)} = S_e(x+y)
\end{aligned}
$$

which implies $NBUC$.

(b) $IMRL \Rightarrow e(x)$ is increasing and so $\lambda_e(x)$ is decreasing. This implies that $f_e(x)$ is DFR and so it is also NWU. This implies $S_e(x)S_e(y) \leq S_e(x+y)$ From (2.21), $S_e(x)/S(x) = e(x)/e(0) \geq 1$ since $e(x)$ is increasing. Therefore, $S_e(x) \geq S(x)$ and so $S_e(x+y) \geq S(x)S_e(y)$ which implies $NWUC$. A similar argument covers the $DMRL$ case.

2.90 (a) $NWUC \Rightarrow S_e(x+y) \geq S_e(y)S(x)$ and with $y = 0, S_e(x) \geq S(x) \Leftrightarrow$ $e(x) \geq e(0)$. Reversing the inequality shows that $NBUC \Rightarrow NBUE$.

(b) From Exercise 2.86, if $e(x) \geq e(0) = E(X)$ then

$$S_e(x) = e^{-\int_0^x [e(y)]^{-1}dy} \geq e^{-\int_0^x [E(X)]^{-1}dy} = e^{-x/E(X)}$$

and reversing the inequality yields the fact that $NBUE \Rightarrow S_e(x) \leq e^{-x/E(X)}$.

(c) In the discussion immediately following (2.22), it was demonstrated that

$$\int_0^\infty S_e(x)dx = \frac{E(X^2)}{2E(X)}.$$

Thus, from (b), in the $NWUE$ case

$$\frac{E(X^2)}{2E(X)} = \int_0^\infty S_e(x)dx \geq \int_0^\infty e^{-x/E(X)dx} = E(X) .$$

That is, $E(X^2) \geq 2[E(X)]^2$ or $Var(X) = E(X^2) - [E(X)]^2 \geq [E(X)]^2$.
Reversing the inequality gives the result for $NBUE$.
(d) Since $e(x) \geq e(0) = E(X)$ from Exercise 2.87,

$$S(x) \leq \frac{E(X)}{x + e(x)} \leq \frac{E(X)}{x + E(X)}$$

(e) From (d),

$$S[E(X)] \leq \frac{E(X)}{E(X) + E(X)} = \frac{1}{2} = S(m)$$

where m is any median. Thus, since $S(x)$ is non-increasing, $E(X) \geq m$.

2.91 (a) $\frac{S_e(x+y)}{S_e(y)} \geq S_e(x)$ may be restated using (2.20) as

$$e^{-\int_y^{x+y}[e(r)]^{-1}dr} \geq e^{-\int_0^x[e(t)]^{-1}dt} .$$

Change variables on the left-hand side from r to $t = r - y$ to obtain

$$e^{-\int_0^x[e(t+y)]^{-1}dt} \geq e^{-\int_0^x[e(t)]^{-1}dt} .$$

Thus, subtracting both sides from 1 and dividing by x gives

$$\frac{1 - e^{-\int_0^x[e(t+y)]^{-1}dt}}{x} \leq \frac{1 - e^{-\int_0^x[e(t)]^{-1}dt}}{x} .$$

(b) Differentiating the numerator and denominator separately yields (by L'Hôpital's rule)

$$\lim_{x\to 0} \frac{1 - e^{-\int_0^x[e(t+y)]^{-1}dt}}{x} = \lim_{x\to 0} \frac{[e(x+y)]^{-1}e^{-\int_0^x[e(t+y)]^{-1}dt}}{1}$$
$$= [e(y)]^{-1}.$$

Similarly, with $y = 0$

$$\lim_{x\to 0} \frac{1 - e^{-\int_0^x[e(t)]^{-1}dt}}{x} = [e(0)]^{-1} .$$

Thus, from (a)

$$\lim_{x\to 0} \frac{1 - e^{-\int_0^x[e(t+y)]^{-1}dt}}{x} \leq \lim_{x\to 0} \frac{1 - e^{-\int_0^x[e(t)]^{-1}dt}}{x}$$

which may be restated as $[e(y)]^{-1} \leq [e(0)]^{-1}$ or $e(y) \geq e(0)$.

(c) From (b), $S_e(x) \geq S(x)$, thus

$$S_e(x+y) \geq S_e(x)S_e(y) \geq S(x)S_e(y)$$

and so $F(x)$ is $NWUC$.

(d) Reverse the inequalities in (a) and (b) to show that $1 - S_e(x)$ is NBU implies that $F(x)$ is $NBUE$. Then reverse the inequalities in (c) to obtain the result.

1.7 SECTION 2.8

2.92 (a)

$$\Pr(\theta = 100 | X = 100) \propto \Pr(X = 100 | \theta = 100) \Pr(\theta = 100) = 0.5\pi(100).$$

$\Pr(\theta = 50 | X = 100) \propto 0.5\pi(50)$. The actual probabilities are

$$\begin{aligned} \Pr(\theta &= 100 | X = 100) = \pi(100)/[\pi(100) + \pi(50)] \text{ and} \\ \Pr(\theta &= 50 | X = 100) = \pi(50)/[\pi(100) + \pi(50)]. \end{aligned}$$

If the switch is not made, the expected outcome is 100. If the switch is made the outcome is 200 if $\theta = 100$ and is 50 if $\theta = 50$. The expected outcome is $\frac{200\pi(100)+50\pi(50)}{\pi(100)+\pi(50)}$.

(b) If switch, the expected outcome is $\frac{200(100)^{-1}+50(50)^{-1}}{100^{-1}+50^{-1}} = 100$.

(c) If the observation is x, switch if $x < \frac{2x\pi(x)+(x/2)\pi(x/2)}{\pi(x)+\pi(x/2)} = \frac{2xe^{-x}+xe^{-x/2}/2}{e^{-x}+e^{-x/2}}$ or $x < 2\log 2$.

2.93 $f(y) = \frac{12(4.801121)^{12}}{y(0.195951+\log y)^{13}}$. Let $W = \log Y - 100 = \log(Y/100)$. Then $y = 100e^w$ and $dy = 100e^w dw$. Thus,

$$f(w) = \frac{12(4.801121)^{12}100e^w}{100e^w(0.195951 + w + \log 100)^{13}} = \frac{12(4.801121)^{12}}{(4.801121 + w)^{13}}, \quad y > 0$$

which is a Pareto density with $\alpha = 12$ and $\theta = 4.801121$.

2.94

$$\begin{aligned} \pi(\alpha|\mathbf{x}) &\propto \frac{\alpha^{10}100^{10\alpha}}{\Pi x_j^{\alpha+1}} \frac{\alpha^{\gamma-1}e^{-\alpha/\theta}}{\theta^\gamma \Gamma(\gamma)} \\ &\propto \alpha^{10+\gamma-1}\exp[-\alpha(\theta^{-1} - 10\log 100 + \Sigma \log x_j)] \end{aligned}$$

which is a gamma distribution with parameters $10+\gamma$ and $(\theta^{-1} - 10\log 100 + \Sigma \log x_j)^{-1}$. The mean is $\hat{\alpha}_{Bayes} = (10 + \gamma)(\theta^{-1} - 10\log 100 + \Sigma \log x_j)^{-1}$. For the *mle*:

$$\begin{aligned} l &= 10\log\alpha + 10\alpha\log 100 - (\alpha+1)\Sigma \log x_j, \\ l' &= 10\alpha^{-1} + 10\log 100 - \Sigma \log x_j = 0 \end{aligned}$$

for $\hat{\alpha}_{mle} = 10(\Sigma \log x_j - 10 \log 100)^{-1}$. The two estimators are equal when $\gamma = 0$ and $\theta = \infty$. This corresponds to $\pi(\alpha) = \alpha^{-1}$, an improper prior.

2.95 Generalizing from Exercise 2.94,

$$\hat{\alpha} = 100(\Sigma \log x_j - 100 \log 100,000)^{-1}$$
$$= 100(1,208.4354 - 100 \log 100,000)^{-1} = 1.75.$$

2.96 (a) $\pi(\mu,\sigma|\mathbf{x}) \propto \sigma^{-n} \exp\left[-\sum \frac{1}{2}\left(\frac{\log x_j - \mu}{\sigma}\right)^2\right]\sigma^{-1}$

(b) Let

$$l = \log \pi(\mu,\sigma|\mathbf{x}) = -(n+1)\log \sigma - \frac{1}{2}\sigma^{-2}\sum(\log x_j - \mu)^2.$$

Then

$$\partial l/\partial \mu = \frac{1}{2}\sigma^{-2}\sum 2(\log x_j - \mu)(-1) = 0$$

and the solution is $\hat{\mu} = \frac{1}{n}\sum \log x_j$. Also,

$$\partial l/\partial \sigma = -(n+1)\sigma^{-1} + \sigma^{-3}\sum(\log x_j - \mu)^2 = 0$$

and so $\hat{\sigma} = \left[\frac{1}{n+1}\sum(\log x_j - \hat{\mu})^2\right]^{1/2}$.

(c)
$$\pi(\mu,\hat{\sigma}|\mathbf{x}) \propto \exp\left[-\sum \frac{1}{2}\left(\frac{\log x_j - \mu}{\hat{\sigma}}\right)^2\right]$$
$$= \exp\left(-\frac{1}{2}\frac{n\mu^2 - 2\mu\Sigma \log x_j + \Sigma(\log x_j)^2}{\hat{\sigma}^2}\right)$$
$$\propto \exp\left(-\frac{1}{2}\frac{\mu^2 - 2\mu\hat{\mu} + \hat{\mu}^2}{\hat{\sigma}^2/n}\right)$$

which is a normal *pdf* with mean $\hat{\mu}$ and variance $\hat{\sigma}^2/n$. The 95% HPD interval is $\hat{\mu} \pm 1.96\hat{\sigma}/\sqrt{n}$.

2.97 (a)
$$\pi(\theta|\mathbf{x}) \propto \frac{(\Pi x_j)\exp(-\theta^{-1}\Sigma x_j)}{\theta^{200}}\frac{\exp(-\lambda\theta^{-1})}{\theta^{\beta+1}}$$
$$\propto \frac{\exp[-\theta^{-1}(\lambda + \Sigma x_j)]}{\theta^{201+\beta}}$$

which is an inverse gamma *pdf* with parameters $200 + \beta$ and $30,000 + \lambda$.
(b) $E(2\theta|\mathbf{x}) = 2\frac{30,000+\lambda}{200+\beta-1}$. At $\beta = \lambda = 0$ it is $2\frac{30,000}{199} = 301.51$ while at $\beta = 2$ and $\lambda = 250$ it is $2\frac{30,250}{201} = 301.00$. For the first case, the inverse gamma parameters are 200 and 30,000. For the lower limit,

$$0.025 = \Pr(2\theta < a) = F(a/2) = 1 - \Gamma(200; 60,000/a)$$

for $a = 262.41$. Similarly the upper limit is 346.34. With parameters 202 and 30,250 the interval is (262.14,345.51).

(c) $Var(2\theta|\mathbf{x}) = 4Var(\theta|\mathbf{x}) = 4\left[\frac{(30,000+\lambda)^2}{(199+\beta)(198+\beta)} - \left(\frac{30,000+\lambda}{199+\beta}\right)^2\right]$. The two variances are 459.13 and 452.99. The two CI's are 301.51 ± 42.00 and 301.00 ± 41.72

(d) $l = -\theta^{-1}30,000 - 200\log\theta$. $l' = \theta^{-2}30,000 - 200\theta^{-1} = 0$. $\hat{\theta} = 150$. For the variance, $l' = \theta^{-2}\Sigma x_j - 200\theta^{-1}$, $l'' = -2\theta^{-3}\Sigma x_j + 200\theta^{-2}$, $E(-l'') = 2\theta^{-3}(200\theta) - 200\theta^{-2} = 200\theta^{-2}$ and so $Var(\hat{\theta}) \doteq \theta^2/200$ and $Var(2\theta) \doteq \theta^2/50$. An approximate CI is $300 \pm 1.96(150)/\sqrt{50} = 300 \pm 41.58$.

1.8   SECTION 2.9

2.98 For the lognormal model, $E(X \wedge 500,000) - E(X \wedge 250,000) = 35,945 - 31,984 = 3,961$. For the Pareto model it is $36,229 - 31,442 = 4,787$.

2.99 The test results are in Table 1.6.

Table 1.6   Calculations for Exercise 2.99

Range	Obs.	Expected	χ^2
0–250	3	$10\left[1 - \left(\frac{1000}{1250}\right)^2\right] = 3.600$	0.100
250–500	2	$10\left[\left(\frac{1000}{1250}\right)^2 - \left(\frac{1000}{1500}\right)^2\right] = 1.956$	0.001
500–1,000	3	$10\left[\left(\frac{1000}{1500}\right)^2 - \left(\frac{1000}{2000}\right)^2\right] = 1.944$	0.574
1,000–	2	$10\left(\frac{1000}{2000}\right)^2 = 2.500$	0.100
Total	10	10.000	0.775

The critical value with 3 degrees of freedom is 6.251. The null hypothesis and therefore the Pareto model is accepted.

2.100 The test results are in Table 1.7.

With two degrees of freedom, the critical value at $\alpha = 0.01$ is 9.210 and at $\alpha = 0.005$ it is 10.597. Accept at $\alpha = 0.005$.

2.101 The *cdf* is $F(x) = \int_0^x (1 + t/2)/2dt = (x + x^2)/2$. The results are in Table 1.8.

The test statistic is 0.320. The critical value is $1.36/\sqrt{5} = 0.608$ and so the null hypothesis and therefore the proposed distribution is accepted.

Table 1.7   Calculations for Exercise 2.100

Range	Obs.	Expected	χ^2
0–3	180	$1,000\left[1-\left(\frac{50}{53}\right)^{3.5}\right]=184.49$	0.109
3–7.5	180	$1,000\left[\left(\frac{50}{53}\right)^{3.5}-\left(\frac{50}{57.5}\right)^{3.5}\right]=202.37$	2.473
7.5–15	235	$1,000\left[\left(\frac{50}{57.5}\right)^{3.5}-\left(\frac{50}{65}\right)^{3.5}\right]=213.93$	2.075
15–40	255	$1,000\left[\left(\frac{50}{65}\right)^{3.5}-\left(\frac{50}{90}\right)^{3.5}\right]=271.40$	0.991
400–	150	$10\left(\frac{50}{90}\right)^{3.5}=127.80$	3.856
Total	1,000	999.99	9.504

Table 1.8   Calculations for Exercise 2.101

Obs.	$F_n(x-)$	$F_n(x)$	$F(x)$	Max. diff.
0.1	0.0	0.2	0.055	0.145
0.4	0.2	0.4	0.280	0.120
0.8	0.4	0.6	0.720	0.320
0.8	0.6	0.8	0.720	0.120
0.9	0.8	1.0	0.855	0.145

2.102 Results for the seven best models, including at least one for each number of parameters are in Table 1.9.

Using the likelihood ratio test, the transformed beta model is superior to the Burr model. It also has the highest p-value for the chi-square test (though the value is not very good). The maximum likelihood estimates are $\hat{\alpha}=2.63295$, $\hat{\theta}=380.506$, $\hat{\gamma}=0.408988$, and $\hat{\tau}=2.72313$.

2.103 There are ten separate samples here. The first sample was of size 42,300 and all that is known is that the lifetime was less than 1.5 for 13 of then. The contribution to the likelihood function is $F(1.5)^{13}[1-F(1.5)]^{42,287}$. Table 1.10 gives the best one, two, and three parameter fits for each of the two mortgage types. Only models that fit both samples with a chi-square test p-value over 0.05 were included.

The likelihood ratio test and the p-values favor the inverse Weibull distribution ($\hat{\tau}=0.600639$ and $\hat{\theta}=49.8671$) for the refinancing model and the inverse gamma distribution ($\hat{\alpha}=0.0783721$ and $\hat{\theta}=4.52473$) for the original mortgages. If a common model is to be used, the total NLL favors the inverse Weibull by a small margin over the inverse gamma. Either model would be reasonable as a common model for the two populations.

Table 1.9   Results for Exercise 2.102

Model	Param.	Negloglike	χ^2	df	p-value
Inv. exponential	1	48,006.11	9,989.74	22	0.0000
Inv. paralogistic	2	44,369.27	51.15	25	0.0015
Lognormal	2	44,381.11	91.39	22	0.0000
Loglogistic	2	44,382.21	66.79	25	0.0000
Burr	3	44,364.34	47.02	23	0.0022
Inv. Burr	3	44,369.09	51.84	24	0.0008
Trans. beta	4	44,359.73	43.18	22	0.0045

Table 1.10   Results for Exercise 2.103

Model	Refinances		Original		Total NLL
	Negloglike	p-value	Negloglike	p-value	
Exponential	680.05	0.000	7,357.30	0.000	8,037.35
Inv. gamma	611.15	0.285	7,291.49	0.670	7,902.64
Inv. Weibull	610.66	0.393	7,291.85	0.524	7,902.51
Burr	610.15	0.372	7,291.88	0.315	7,902.03
Gen. Pareto	610.54	0.243	7,291.45	0.481	7,901.99

2.104 The best one parameter model in terms of NLL and chi-square test statistic is the exponential with 7,744.94 and 337.30 respectively. The best two parameter model is the gamma with 7,716.40 and 276.65. By the By the likelihood ratio test the gamma model is a significant improvement.. The best three parameter model is the inverse Burr with 7,714.19 and 269.78. Although the likelihood ratio test cannot be used, the improvement over the gamma (4.42 after doubling) is significant ($p = 0.0355$ with one degree of freedom), given the small number of groups, the simpler two parameter model will probably suffice. The gamma parameters are $\hat{\alpha} = 0.776683$ and $\hat{\theta} = 43,136.6$ while the Burr parameters are $\hat{\tau} = 0.210272$, $\hat{\theta} = 69,228.3$, and $\hat{\gamma} = 2.67014$.

2.105 The results for the K-S test are in Table 1.11.

The test statistic is 0.172 and even at $\alpha = 0.2$ where the critical value is $1.07/\sqrt{10} = 0.338$, the null hypothesis, and therefore the gamma distribution, is accepted.

2.106 The results for the K-S test are in Table 1.12.

Table 1.11   Results for Exercise 2.105

Obs.	$F_n(x-)$	$F_n(x)$	$F(x)$	Max. diff.
1,500	0.0	0.1	0.030	0.070
1,800	0.1	0.2	0.063	0.137
3,000	0.2	0.3	0.331	0.131
3,500	0.3	0.4	0.472	0.172
3,800	0.4	0.5	0.553	0.153
3,900	0.5	0.6	0.579	0.079
4,200	0.6	0.7	0.651	0.051
4,800	0.7	0.8	0.771	0.071
5,500	0.8	0.9	0.869	0.069
6,000	0.9	1.0	0.915	0.085

The value of the test statistic is 0.143 and even at $\alpha = 0.2$ where the critical value is $1.07/\sqrt{20} = 0.239$, the null hypothesis, and therefore the single parameter Pareto distribution, is accepted.

2.107 The best one parameter model is the inverse exponential with $NLL = 659.76$ and a p-value of 0.3241. The best two parameter model is the inverse gamma with $NLL = 659.28$ and a p-value of 0.3528. While the p-value has improved, the improvement in the NLL is not enough to warrant the second parameter (twice the difference is 0.96 which does not exceed the 3.84 cutoff at 5% significance with one degree of freedom). The inverse transformed gamma distribution has $NLL = 658.68$ with a p-value of 0.3397. Two extra parameters cannot be justified as twice the improvement in NLL is 2.16 which does not exceed the 5% cut-off of 5.99. The inverse exponential is the best model for this data.

2.108 Table 1.13 gives information about those models that fit well to both data sets. In terms of NLL, the best for each data set are in the list.

For the 100,000 limit, the inverse Pareto has the best NLL and the inverse Burr is an insignificant improvement. For the 1,000,000 limit the same holds for the Pareto and inverse Burr. While different models were selected for each limit, the lowest total NLL occurs with the inverse Pareto distribution. It is an acceptable second choice for the 1,000,000 limit and so a common distribution is viable. When the parameters are forced to match, the NLL is 1,042.89. The increase is only 0.29. When doubled, 0.58 is not significant versus a chi-square distribution with two degrees of freedom. A model with common inverse Pareto parameters ($\hat{\tau} = 0.892218$ and $\hat{\theta} = 15,071.8$) is appropriate.

Table 1.12 Results for Exercise 2.106

Obs.	$F_n(x-)$	$F_n(x)$	$F(x)$	Max. diff.
102	0.00	0.05	0.055	0.055
102	0.05	0.10	0.055	0.045
107	0.10	0.15	0.175	0.075
108	0.15	0.20	0.197	0.047
108	0.20	0.25	0.197	0.053
110	0.25	0.30	0.238	0.062
110	0.30	0.35	0.238	0.112
111	0.35	0.40	0.257	0.143
117	0.40	0.45	0.361	0.089
132	0.45	0.50	0.546	0.096
135	0.50	0.55	0.575	0.075
135	0.55	0.60	0.575	0.025
147	0.60	0.65	0.666	0.066
147	0.65	0.70	0.666	0.034
149	0.70	0.75	0.679	0.071
165	0.75	0.80	0.760	0.040
176	0.80	0.85	0.800	0.050
226	0.85	0.90	0.902	0.052
227	0.90	0.95	0.903	0.047
476	0.95	1.00	0.988	0.038

2.109 Table 1.14 provides some fits for the two limits.

The blank cells refer to cases where the *mle* does not exist. For the 10,000 limit the transformed gamma is slightly better than the Weibull. Twice the difference in the NLL's is 3.92 which is barely significant at 5%. For the 25,000 limit the Burr is better than the lognormal. The total NLL is 2,309.17 + 4,255.11 = 6,564.28. When a common model is used, the Burr is significantly better than the lognormal (twice the difference is 12.36 and there are two degrees of freedom). The common (Burr) model adds only 0.77 to the NLL so it is reasonable to use a common model. The Burr is not a bad second choice to the transformed gamma for the 10,000 limit.

2.110 The results of some of the better fits are in Table 1.15.

The likelihood ratio test favors the Weibull over the exponential. It also favors the transformed gamma over the Weibull as twice the difference in NLL's is 27.34. The chi-square test rejects all of these models, but looks more favorably on the transformed gamma ($p = 3.22 \times 10^{-24}$) than the Weibull ($p = 2.27 \times 10^{-27}$). Use the transformed gamma distribution.

Table 1.13   Results for Exercise 2.108

Model	100,000 limit		1,000,000 limit		
	Negloglike	p-value	Negloglike	p-value	Total NLL
Pareto	222.04	0.117	820.78	0.526	1,042.82
Inv. Pareto	221.79	0.120	820.81	0.516	1,042.60
Loglogistic	222.05	0.103	820.97	0.492	1,043.02
Paralogistic	222.07	0.104	820.94	0.496	1,043.01
Inv. Paralogistic	222.01	0.104	820.96	0.493	1,042.97
Burr	221.48	0.116	820.25	0.495	1,041.73
Inv. Burr	221.19	0.155	820.16	0.533	1,041.35
Gen. Pareto	221.35	0.130	820.19	0.525	1,041.54

Table 1.14   Results for Exercise 2.109

Model	10,000 limit		25,000 limit		
	Negloglike	p-value	Negloglike	p-value	Total NLL
Lognormal	2,313.59	0.000	4,257.64	0.463	6,571.23
Inv. Pareto	2,315.60	0.000	4,257.68	0.446	6,573.28
Weibull	2,311.13	0.000	4,278.05	0.000	6,589.18
Burr	2,309.94	0.000	4,255.11	0.879	6,565.05
Trans. gamma	2,309.17	0.000	–	–	
Trans. beta	–	–	4,255.11	0.809	

2.111 The SBC for the model with common parameters is

$$54,260.96 + 2\log(26,655/2\pi) = 54,277.67.$$

With a common τ it is

$$54,079.38 + 6\log(26,655/2\pi) = 54,129.50.$$

With distinct parameters it is

$$54,047.52 + 10\log(26,655/2\pi) = 54,131.05.$$

The smallest value is for the model with common τ but distinct θ.

1.9   SECTION 2.10

2.112 The *mle* is $\hat{\tau} = 1.04238$ and $\hat{\theta} = 49.1462$. The covariance matrix is $\begin{bmatrix} .005196265 & -0.2249201 \\ -0.2249201 & 14.68333 \end{bmatrix}$. The expected number of unreported claims is $463[e^{(\theta/168)^{\tau}} - 1]$. The point estimate is 148.19.

Table 1.15 Results for Exercise 2.110

Model	negloglike	χ^2	df
Exponential	4946.59	304.66	24
Loglogistic	4,937.78	329.22	23
Gamma	4,892.57	204.45	23
Weibull	4,881.48	185.48	23
Inv. Burr	4,868.48	169.55	22
Trans. gamma	4,867.81	166.69	22

The derivatives are (evaluated at the estimated values) with respect to τ: $463e^{(\theta/168)^\tau}(\theta/168)^\tau \log(\theta/168) = -208.616$, and with respect to θ: $463e^{(\theta/168)^\tau}(\theta/168)^\tau \tau/\theta = 3.59976$.

The variance of the point estimator is 754.231 and so the confidence interval is $148.19 \pm 2\sqrt{754.231}$ which is 148.19 ± 54.93.

2.113 When shifted at 75 the effective deductible is 25 which must be added to each payment. The 25th and 50th percentiles are 361.25 and 925 and so the percentile matching equations are

$$
\begin{aligned}
0.25 &= F_Y(361.25) = \frac{F_X(361.25) - F_X(25)}{1 - F_X(25)} \\
&= \frac{\exp[-(25/\theta)^\tau] - \exp[-(361.25/\theta)^\tau]}{\exp[-(25/\theta)^\tau]} \\
0.50 &= F_Y(925) = \frac{F_X(925) - F_X(25)}{1 - F_X(25)} \\
&= \frac{\exp[-(25/\theta)^\tau] - \exp[-(925/\theta)^\tau]}{\exp[-(25/\theta)^\tau]}.
\end{aligned}
$$

The solution is $\hat{\tau} = 0.87308$ and $\hat{\theta} = 1,338.8$.

The five smaller observations contribute $f_X(x + 25)/[1 - F_X(25)]$ to the likelihood where x is the observed payment. The five observations at the limit each contribute $[1 - F_X(925)]/[1 - F_X(25)]$. The likelihood is maximized at $\hat{\tau} = 0.778187$ and $\hat{\theta} = 1,343.67$.

2.114

$$
\begin{aligned}
E(Y^*) &= \frac{22}{15}\int_{100}^{250}(x - 100)f(x)dx + \frac{22}{25}\int_{250}^{1800} xf(x)dx \\
&\quad + 1584[1 - F(1800)] \\
&= \frac{22}{15}\int_0^{250} xf(x)dx - \frac{22}{15}\int_0^{100} xf(x)dx - \frac{2200}{15}[F(250) - F(100)]
\end{aligned}
$$

$$+\frac{22}{25}\int_0^{1800} xf(x)dx - \frac{22}{25}\int_0^{250} xf(x)dx + 1584[(1-F(1800)]$$

$$= \frac{22}{15}E(X \wedge 250) - \frac{22(250)}{15}[1-F(250)] - \frac{22}{15}E(X \wedge 100)$$

$$+\frac{22(100)}{15}[1-F(100)] - \frac{2200}{15}[F(250)-F(100)]$$

$$+\frac{22}{25}E(X \wedge 1800) - \frac{22(1800)}{25}[1-F(1800)]$$

$$-\frac{22}{25}E(X \wedge 250) + \frac{22(250)}{25}[1-F(250)]$$

$$+1584[1-F(1800)]$$

$$= \frac{22}{25}E(X \wedge 1800) + \frac{44}{74}E(X \wedge 250) - \frac{22}{15}E(X \wedge 100)$$

2.115 (a) $\Pr(Loss > 25) = \Pr(X > 25 | X > 10) = \dfrac{1-F(25)}{1-F(10)}$

$$= \frac{\exp(-25/20)}{\exp(-10/20)} = 0.4724.$$

(b) $\Pr(Y < 25) = \Pr(X < 35 | X > 10) = \dfrac{F(35)-F(10)}{1-F(10)}$

$$= \frac{\exp(-10/20) - \exp(-35/10)}{\exp(-10/20)} = 0.7135.$$

2.116 $\Pr(P < 5) = \Pr(d < 1.1X < d+5 | 1.1X > d)$

$$= \Pr[d/1.1 < X < (d+5)/1.1 | X > d/1.1]$$

$$= [-e^{-.1(d+5)/1.1} + e^{-.1(d)/1.1}]/e^{-.1(d)/1.1}$$

$$= 1 - e^{-.1(5)/1.1} = .3653$$

2.117 $F(x) = 1 - e^{-\lambda x}$.

$$L(\lambda) = \frac{F(2,000)^2[F(5,000)-F(2,000)]^4}{F(5,000)^6}$$

$$= \frac{(1-e^{-2,000\lambda})^2(e^{-2,000\lambda}-e^{-5,000\lambda})^4}{(1-e^{-5,000\lambda})^6}.$$

2.118 (a) $10\int_{100}^{\infty} x(2)1,000^2(x+1,000)^{-3}dx$. Let $y = x+1000$. Then we have

$$10\int_{1100}^{\infty} (y-1,000)2(1,000)^2 y^{-3}dy = 20(1,000)^2 \left.\frac{y^{-1}}{-1} - 1,000\frac{y^{-2}}{-2}\right|_{1100}^{\infty}$$

$$= 9M917.36.$$

(b)
$$10 \int_{100/1.1}^{\infty} 1.1x(2)1,000^2(x+1,000)^{-3}dx$$
$$= 22(1,000)^2 \int_{1090.91}^{\infty} (y-1,000)2(1,000)^2 y^{-3}dy$$
$$= 10,923.61.$$

2.119 With regard to the negative loglikelihood, the best fits are: 1 parameter – inverse exponential, 456.940, 2 parameter – inverse Gaussian, 454.012, 3 parameter – transformed gamma, 454.091. Although the inverse exponential is not a special case of the inverse Gaussian, the improvement of 2.828 justifies the additional parameter. As no three parameter model had a lower NLL than the inverse Gaussian, it is the best choice. The parameter estimates are $\hat{\mu} = 203,500$ and $\hat{\theta} = 38,060.3$.

2.120 The best model is Weibull with $\hat{\tau} = 0.153691$ and $\hat{\theta} = 0.452029$. This model easily passes the chi-square goodness-of-fit test with a p-value of 0.9219. Convergence was not obtained for any three parameter models and none had a smaller negative loglikelihood. The expected payment per payment is

$$\frac{E(X \wedge 1,000,000) - E(X \wedge 5,000)}{1 - F(5,000)} = \frac{747.78 - 154.47}{1 - .984749} = 38,903.$$

2.121 The best model is the lognormal with $\hat{\mu} = 11.0447$ and $\hat{\sigma} = 1.65859$. No other fitted model had a smaller negative loglikelihood. For the truncated model in Exercise 2.117 the expected payment (in thousands) is

$$\frac{E(X) - E(X \wedge 25,000)}{1 - F(25,000)} = \frac{203,500 - 22,357}{1 - .260155} = 244,839$$

while for the shifted model from this problem the expected payment per payment (in thousands) is

$$\frac{E(X) - E(X \wedge 20,000)}{1 - F(20,000)} = \frac{247,747 - 17,432}{1 - 0.245706} = 305,339.$$

2.122 The best candidates with one, two, and three parameters are the exponential ($NLL = 2,910.77$), inverse Weibull ($NLL = 2,904.34$), and inverse transformed gamma ($NLL = 2,902.49$) respectively. All have very high chi-square goodness of fit test statistics. The likelihood improvement warrants consideration of the inverse Weibull, but the likelihood ratio test rejects the inverse transformed gamma. Choose the inverse Weibull with $\hat{\tau} = 2.20088$ and $\hat{\theta} = 41,896.9$.

2.123 Nine of the data points are eliminated. The eight which are below 200 (yet above 125) each contribute $f(x)/[1 - F(125)]$ or $\alpha(125)^\alpha x^{-\alpha-1}$ to the likelihood function. The three observations at the limit each contribute $[1 - F(200)]/[1 - F(125)]$ which is $(125/200)^\alpha$. Then

$$L = \alpha^8 (125)^{8\alpha} (\textstyle\prod x_i)^{-\alpha-1} (125/200)^{3\alpha}$$

$$
\begin{aligned}
l &= 8\log a + 11\alpha \log 125 - 3\alpha \log 200 - (\alpha + 1)(39.95459) \\
&= 8\log \alpha - 2.73809\alpha.
\end{aligned}
$$

The maximum occurs at $\hat{\alpha} = 2.9217$. The answer is similar to the value of 2.848 which was obtained when no observations were truncated or censored.

2.124 The best fitting distributions are inverse exponential ($NLL = 164.89$ and a p-value of 0.2522), Pareto ($NLL = 164.51$ and a p-value of 0.1734), and inverse Weibull ($NLL = 164.52$ and a p-value of 0.1719). There is no justification for the two parameter models and so choose the inverse exponential distribution with $\hat{\theta} = 203.485$. The requested value is $E(X \wedge 2500) = 604.63$. For the inverse exponential model of Exercise 2.107 with $\hat{\theta} = 93.1877$ we have $E(X \wedge 2500) = 347.66$.

2.125 The best models of one through three parameters are exponential, Weibull, and transformed gamma with negative logliklihoods of 4,246.78, 4,192.41 and 4,189.57. The likelihood ratio test rejects the exponential in favor of the Weibull and then rejects the Weibull in favor of the transformed gamma. The parameter estimates are $\hat{\alpha} = 0.386573$, $\hat{\theta} = 90.0735$, and $\hat{\tau} = 2.83224$.

The expected payment per loss is

$$
\begin{aligned}
\int_d^\infty x f(x)dx &= \int_0^\infty x f(x)dx - \int_0^d x f(x)dx \\
&= E(X) - E(X \wedge d) + d[1 - F(d)]
\end{aligned}
$$

Dividing by $1 - F(d)$, the probability of exceeding the deductible we obtain the expected payment per payment of

$$d + \frac{E(X) - E(X \wedge d)}{1 - F(d)} = 6 + \frac{48.596 - 5.834}{1 - 0.058001} = 51.395.$$

1.10   SECTION 2.11

2.126 $Var(X_K) = \hat{\mu}_2 + \left(\frac{\delta}{\mu^2} - 1\right)\hat{\mu}_2'$. For the Pareto distribution, $\mu = (\alpha - 1)^{-1}$, $\delta = E(X^2)/\theta^2 = 2[(\alpha - 1)(\alpha - 2)]^{-1}$ and so $Var(X_K) = \hat{\mu}_2 + \frac{\alpha}{\alpha-2}\hat{\mu}_2'$.

2.127 Let $\boldsymbol{\alpha}$ be the parameters of the marginal distribution of X. Then from the marginal distribution we have $L(\boldsymbol{\alpha}; \mathbf{x}) = \prod f_X(x_i; \boldsymbol{\alpha})$. The joint density function is $f_{X,Y}(x, y) = f_{Y|X}(y; \boldsymbol{\beta}, x) f_X(x; \boldsymbol{\alpha})$ and so $L(\boldsymbol{\alpha}, \boldsymbol{\beta}; \mathbf{x}, \mathbf{y}) = \prod f_{Y|X}(y; \boldsymbol{\beta}, x) \prod f_X(x_i; \boldsymbol{\alpha})$. The value of $\boldsymbol{\alpha}$ that maximizes one of these likelihoods must maximize the other.

2.128 We have $f_K(x; \alpha) = \frac{1}{2} \sum_{j=1}^{n} \frac{\alpha[(\alpha-1)x_j]^{\alpha}}{[x+(\alpha-1)x_j]^{\alpha+1}}$. This function is plotted for $\alpha = 3$ in Figure 1.4

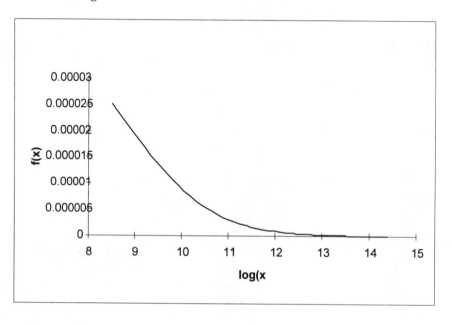

Fig. 1.4   Kernel density estimator for Exercise 2.128

2.129 (a) $f(x) = \theta x^{-2} e^{-\theta/x}$, $f(y|x) = (\gamma + \beta \log x)^{-1} e^{-y(\gamma + \beta \log x)^{-1}}$,

$$f(x, y) = \frac{\theta e^{-\theta/x - y(\gamma + \beta \log x)^{-1}}}{x^2(\gamma + \beta \log x)}.$$

(b)

$$\int_0^{\infty} \int_0^{50,000} (x + y) \frac{\theta e^{-\theta/x - y(\gamma + \beta \log x)^{-1}}}{x^2(\gamma + \beta \log x)} dx dy$$

$$= \int_0^{50,000} \frac{\theta e^{-\theta/x}}{x^2} \int_0^{\infty} \frac{(x + y) e^{-y(\gamma + \beta \log x)^{-1}}}{(\gamma + \beta \log x)} dy dx$$

$$= \int_0^{50,000} \frac{\theta e^{-\theta/x}}{x^2} \left[-(x + y) e^{-y(\gamma + \beta \log x)^{-1}} \right.$$

$$-(\gamma + \beta \log x)e^{-y(\gamma + \beta \log x)^{-1}} \Big] \Big|_0^\infty dx$$

$$= \int_0^{50,000} \frac{\theta e^{-\theta/x}}{x^2}[x + (\gamma + \beta \log x)]dx.$$

(c) The maximization was done using the EXCEL solver. The solution was $\hat{\theta} = 6{,}352$, $\hat{\gamma} = 0.004820$, and $\hat{\beta} = 564.4$.

2.130 Separate fitting of the marginal distributions produces scale parameter estimates of 6,365 and 216.2 for the loss and ALAE respectively. The combined NLL is 496.58. Fitting Frank's copula produces scale parameter estimates of 5,886 and 218.2 as well as $\hat{\alpha} = 0.3035$. The NLL is 495.81. With one degree of freedom, the reduction in the NLL of 0.77 is not significant. Assuming the inverse exponential distribution is correct, the association is not significant.

Chapter 3 Solutions

2.1 SECTION 3.1

3.1

$$P_N^{(k)}(z) = \sum_{j=0}^{\infty} p_j \frac{d^k}{dz^k} z^j = \sum_{j=k}^{\infty} p_j j(j-1)\cdots(j-k+1)z^{j-k},$$

$$P_N^{(k)}(0) = p_k k!.$$

3.2

$$P_N(z) = \sum_{k=0}^{\infty} p_k z^k = \sum_{k=0}^{\infty} p_k e^{k\log z} = M_N(\log z)$$

$$P_N'(z) = z^{-1} M_N'(\log z)$$

$$P_N'(1) = M_N'(0) = E(N)$$

$$P_N''(z) = -z^{-2} M_N'(\log z) + z^{-2} M_N''(\log z)$$

$$P_N''(1) = -E(N) + E(N^2) = E[N(N-1)]$$

2.2 SECTION 3.2

3.3 (a) The *mle* is the sample mean, $[905 + 2(45) + 3(2)]/10{,}000 = 0.1001$.
(b) $0.1001 \pm 1.96\sqrt{0.1001/10{,}000} = 0.1001 \pm 0.0062$ or $(0.0939, 0.1063)$.
(c) The expected counts are:

$$0 \ : \ 10{,}000(e^{-0.1001}) = 9{,}047.47,$$

$$1 \ : \ 10{,}000(0.1001e^{-0.1001}) = 905.65,$$

$$2 \ : \ 10{,}000(0.1001^2 e^{-0.1001}/2) = 45.33,$$

$$3 \text{ or more} \ : \ 10{,}000 - 9{,}047.47 - 905.65 - 45.33 = 1.55.$$

Combine the last two groups for 2 or more: 46.88. The test statistic is

$$(9{,}048 - 9{,}047.47)^2/9{,}047.47 + (905 - 905.65)^2/906.65$$
$$+(47 - 46.88)^2/46.88 = 0.0008.$$

There is one degree of freedom (three groups less one and less one estimated parameter) and so the 5% critical value is 3.84 and the null hypothesis (and therefore the Poisson model) is accepted.

3.4
$$0.1001 + 1.9208/10{,}000 \pm 0.5\sqrt{\frac{15.3664(0.1001) + 3.8416^2/10{,}000}{10{,}000}}$$

$$= 0.1003 \pm 0.0062$$

or $(0.0941, 0.1065)$.

3.5 (a) The sample means are – underinsured: $109/1{,}000 = 0.109$ and insured: $57/1{,}000 = 0.057$.

(b) For underinsured the expected counts are 896.73, 97.74, and 5.53. The test statistic is $0.02 + 0.34 + 0.39 = 0.75$. With one degree of freedom, the 5% critical value is 3.84 and the null hypothesis (and therefore the Poisson model) is accepted. For uninsured the expected counts are 944.59, 53.84, and 1.57. In order to achieve the minimum expected count of 5, the last two groups must be combined to yield an observed of 53 and an expected of 55.41. While the test statistic can be computed (0.11) there are zero degrees of freedom, so the test cannot be completed.

(c) The Poisson parameter is the sum of the individual parameters, $0.109 + 0.057 = 0.166$.

3.6 (a) The sample mean is $166/1{,}000 = 0.166$.

(b) Let n_{ij} be the number observations of j counts from population i where $j = 0, 1, \ldots$ and $i = 1, 2$. The individual estimators are $\hat{\lambda}_i = \sum_{j=0}^{\infty} j n_{ij}$. From the Theorem the estimator for the sum is the sum of the estimators which is $\hat{\lambda} = \sum_{j=0}^{\infty} j(n_{1j} + n_{2j})$ which is also the estimator from the combined sample.

(c) The expected counts are 847.05, 140.61, 11.67, and 0.67. Combining the last two groups gives the test statistic $0.23 + 0.73 + 0.60 = .56$. This exceeds the critical value for one degree of freedom (3.84) and so the null hypothesis (and therefore the Poisson model) is rejected.

(d) If they were independent, N would also have a Poisson distribution. It does not and so the independence hypothesis seems unlikely. This is plausible because the circumstances that lead one to have accidents involving under-insured motorists may be similar to those that lead to accidents involving uninsured motorists.

3.7 (a) $\hat{\lambda} = 15.688$. The *NLL* is 3,578.58.
(b) $l''(15.688) = 29.863$, $Var(\hat{\lambda}) \doteq 1/29.863 = 0.033486$. The CI is $15.688 \pm 1.96(0.18299)$ or 15.688 ± 0.359.
(c) The calculations are in Table 2.1. There are three degrees of freedom. The Poisson model is rejected.

Table 2.1 Calculations for Exercise 3.7

Interval	Obs.	Exp.	Chi-square
0–10	225	44.60	729.67
11–15	57	205.87	107.66
16–20	40	194.64	122.86
21–25	38	52.61	4.06
26–	143	5.28	3,594.15
Total			4,558.40

2.3   SECTION 3.3

3.8 (a) $r\beta = 0.1001$, $r\beta(1 + \beta) = 0.1103 - 0.1001^2 = 0.10027999$. $1 + \beta = 1.0017981$, $\hat{\beta} = 0.0017981$, $\hat{r} = 55.670$.
(b) Even with an expected of 1.63, the last group is retained, else there would not be enough degrees of freedom. There is 1 *df* and so the 5% critical value is 3.84 and the negative binomial model is accepted. The calculations are in Table 2.2.

Table 2.2 Calculations for Exercise 3.8

Value	Obs.	Exp.	Chi-square
0	9,048	9,048.28	0.000
1	905	904.11	0.001
2	45	45.98	0.021
3+	2	1.63	0.084
Total			0.106

(c)

$$H(\hat{r}) = 10{,}000 \log\left(1 + \frac{0.1001}{\hat{r}}\right) - 905\left(\frac{1}{\hat{r}}\right)$$
$$-45\left(\frac{1}{\hat{r}} + \frac{1}{\hat{r}+1}\right) - 2\left(\frac{1}{\hat{r}} + \frac{1}{\hat{r}+1} + \frac{1}{\hat{r}+2}\right).$$

It is zero at $\hat{r} = 56.1872$ and then $\hat{\beta} = 0.1001/56.1872 = 0.00178154$.

(d) For the Poisson distribution $\hat{\lambda} = 0.1001$ and the *NLL* is 3,339.66. With the parameters in (c) the negative binomial *NLL* is 3,339.65. The test statistic (twice the difference) is 0.02 which is not significant (the 5% critical value is 3.84). Use the Poisson model.

3.9 (a) $r\beta = 0.166$, $r\beta(1+\beta) = 0.252 - 0.166^2 = 0.224444$. $1 + \beta = 1.352072$, $\hat{\beta} = 0.352072$, $\hat{r} = 0.47149$.

(b) With one degree of freedom, the critical value is 3.84 and the negative binomial distribution is rejected. See Table 2.3.

Table 2.3   Calculations for Exercise 3.9

Value	Obs.	Exp.	Chi-square
0	861	867.43	0.048
1	121	106.50	1.974
2	13	20.40	2.684
3+	5	5.67	0.079
Total			4.785

(c) $\hat{r} = 0.656060$ and $\hat{\beta} = 0.253026$.

(d) The Poisson *NLL* is 488.241 and the negative binomial *NLL* is 476.457. The test statistic is $2(488.241 - 476.457) = 23.568$ while the critical value is 3.84. Reject the Poisson in favor of the negative binomial.

(e) Yes, it could be a mixture of different independent Poissons.

(f) The calculations appear in Table 2.4. The zero totals in columns (6) and (7) indicate that the maximum was achieved. The information matrix is 1,000 (the sample size) times the last three column totals:

$$\begin{bmatrix} 328.738 & 798.067 \\ 798.067 & 2{,}069.260 \end{bmatrix}$$

and its inverse is

$$\begin{bmatrix} 0.047754 & -0.018418 \\ -0.018418 & 0.0075865 \end{bmatrix}.$$

(g) For r: $0.65606 \pm 1.96\sqrt{0.047754}$ or 0.65606 ± 0.42831. For β: $0.25303 \pm 1.96\sqrt{0.0075865}$ or 0.25303 ± 0.17072.

Table 2.4 Calculations for Exercise 3.9

(1)	(2)	(3)	(4)	(5)	(6)	(7)	(8)	(9)	(10)
k	n_k	p_k	$\partial p_k/\partial r$	$\partial p_k/\partial \beta$	$\dfrac{(2)(4)}{(3)}$	$\dfrac{(2)(5)}{(3)}$	$\dfrac{(4)(4)}{(3)}$	$\dfrac{(4)(5)}{(3)}$	$\dfrac{(5)(5)}{(3)}$
0	861	.8624	−.1945	−.4516	−194.21	−450.80	.0439	.1019	.2364
1	121	.1143	.1484	.3006	157.14	318.29	.1927	.3903	.7906
2	13	.0191	.0364	.1105	24.73	75.20	.0692	.2103	.6393
3	3	.0034	.0078	.0305	6.84	26.82	.0177	.0696	.2729
4	1	.0006	.0016	.0076	2.55	12.09	.0041	.0195	.0922
5	0	.0001	.0003	.0018	0.00	0.00	.0009	.0050	.0276
6	1	.0000	.0001	.0004	2.94	18.40	.0002	.0012	.0076
7+	0	.0000	.0000	.0001	0.00	0.00	.0001	.0004	.0027
Tot.		1.0000	.0000	.0000	0.00	0.00	.3287	.7981	2.0693

3.10 $\alpha = r = 0.480689$ and $\theta = \beta = 0.349886$.

3.11 Because $r = 1$, $\hat{\beta} = \bar{X}$.

$$Var(\bar{X}) = Var(X)/n = \beta(1+\beta)/n.$$

$$l = \sum_{j=1}^{n} \log \Pr(N = x_j) = \sum_{j=1}^{n} \log[\beta^{x_j}(1+\beta)^{-x_j-1}]$$

$$= \sum_{j=1}^{n} x_j \log(\beta) - (x_j + 1)\log(1+\beta)$$

$$l'' = \sum_{j=1}^{n} -x_j\beta^{-2} + (x_j + 1)(1+\beta)^{-2}$$

$$E(l'') = -n\beta\beta^{-2} + n(\beta + 1)(1+\beta)^{-2} = n/[\beta(1+\beta)]$$

The reciprocal matches the true variance of the *mle*.

3.12 For Exercise 3.8: (a) $\hat{\beta} = \bar{x} = 0.1001$.
(b) $Var(\hat{\beta}) \doteq 0.1001(1.1001)/10{,}000 = 0.000011012$.
(c) With two degrees of freedom, the critical value is 5.99 and the geometric model is rejected. See Table 2.5.

Table 2.5   Calculations for Exercise 3.12

Value	Obs.	Exp.	Chi-square
0	9,048	9,090.08	0.195
1	905	827.12	7.333
2	45	75.26	12.167
3+	2	7.54	4.071
Total			23.766

(d) The *NLL* is 3,353.40 while for the negative binomial model it is 3,339.65. Twice the difference is 27.50 which exceeds 3.84. The negative binomial model is superior to the geometric.
(e) From Exercise 3.8 the negative binomial is not superior to the Poisson. Use the Poisson distribution.
     For Exercise 3.9: (a) $\hat{\beta} = \bar{x} = 0.166$.
(b) $Var(\hat{\beta}) \doteq 0.166(1.166)/1{,}000 = 0.000193556$.
(c) With two degrees of freedom, the critical value is 5.99 and so the geometric model is accepted. See Table 2.6
(d) The geometric *NLL* is 477.171 while for the negative binomial it is 476.457. Twice the difference is 1.428 which is not significant when compared to 3.841. Choose the geometric model.

Table 2.6   Calculations for Exercise 3.12

Value	Obs.	Exp.	Chi-square
0	861	844.90	0.307
1	121	121.64	0.003
2	13	17.51	1.162
3+	5	2.95	1.425
Total			2.897

(e) The Poisson *NLL* is 488.241 and so the one-parameter geometric model is preferred.

3.13 (a) geometric - $\hat{\beta} = 19.146$, negative binomial - $\hat{r} = 0.564181$, $\hat{\beta} = 37.9032$.
(b) The calculations appear in Table 2.7. For the geometric there are 7 *df* and the 5% critical value is 14.07 and so the model is rejected. For the negative binomial there are 6 *df* and the 5% critical value is 12.59 and so that model is also rejected.

Table 2.7   Calculations for Exercise 3.13

Value	Obs.	Geometric Exp.	Geometric Chi-square	Negative binomial Exp.	Negative binomial Chi-Square
0	82	25.0	130.26	63.8	5.22
1–3	49	67.7	5.17	84.0	14.58
4–6	47	58.1	2.13	52.0	0.48
7–10	47	64.9	4.93	50.6	0.26
11–15	57	64.6	0.89	46.9	2.16
16–20	40	50.1	2.02	35.8	0.50
21–25	38	38.8	0.02	28.2	3.37
26–35	52	53.4	0.04	41.4	2.73
36+	91	80.5	1.38	100.3	0.87
Total			146.84		30.16

(c) For Poisson vs. geometric the *NLL*'s are 3,578.58 and 1,132.25 respectively and so the geometric is preferred. For the negative binomial it is 1,098.64. Twice the difference (vs. geometric) is 67.22 which is clearly significant. Choose the negative binomial.
(d) From the scoring method the variance is estimated as 0.917849. The CI is $19.145 \pm 1.96\sqrt{0.917849}$ or 19.145 ± 1.878.

(e) The expected benefit is

$$0\frac{1}{1+\beta} + 10\frac{\beta}{(1+\beta)^2} + 20\frac{\beta^2}{(1+\beta)^3} + \cdots$$
$$+190\frac{\beta^{19}}{(1+\beta)^{20}} + 200\frac{\beta^{20}}{(1+\beta)^{20}}$$

which is $10\left[\beta - \frac{\beta^{21}}{(1+\beta)^{20}}\right]$. At $\beta = 19.145$ it is 122.30. The derivative is

$$10\left[1 - \frac{(1+\beta)^{20}21\beta^{20} - \beta^{21}20(1+\beta)^{19}}{(1+\beta)^{40}}\right]$$

which, when evaluated at $\beta = 19.145$, is 2.80176. The CI is

$$122.30 \pm 1.96(2.80176)\sqrt{0.917849} \text{ or } 122.30 \pm 5.26.$$

2.4 SECTION 3.4

3.14 (a) $\hat{q} = \bar{X}/m$, $E(\hat{q}) = E(X)/m = mq/m = q$.
(b) $\quad\quad Var(\hat{q}) \; = \; Var(\bar{X})/m^2 = Var(X)/(nm^2)$
$$= \; mq(1-q)/(nm^2) = q(1-q)/(nm).$$

(c)
$$l \; = \; \sum_{j=1}^{n}\log\binom{m}{x_j} + x_j\log q + (m-x_j)\log(1-q)$$
$$l' \; = \; \sum_{j=1}^{n} x_j q^{-1} - (m-x_j)(1-q)^{-1}$$
$$l'' \; = \; \sum_{j=1}^{n} -x_j q^{-2} - (m-x_j)(1-q)^{-2}$$
$$I(q) \; = \; E(-l'') = n[mqq^{-2} + (m-mq)(1-q)^{-2}]$$
$$= \; nm[q^{-1} + (1-q)^{-1}].$$

The reciprocal is $(1-q)q/(nm)$.
(d) $\hat{q} \pm z_{\alpha/2}\sqrt{\hat{q}(1-\hat{q})/(nm)}$.
(e) $\quad\quad 1 - \alpha = \Pr\left(-z_{\alpha/2} \le \frac{\hat{q}-q}{\sqrt{q(1-q)/(nm)}} \le z_{\alpha/2}\right)$

and so

$$|\hat{q} - q| \le z_{\alpha/2}\sqrt{\frac{q(1-q)}{nm}}$$

which implies

$$nm(\hat{q} - q)^2 \le z_{\alpha/2}^2 q(1-q).$$

Then

$$(nm + z_{\alpha/2}^2)q^2 - (2nm\hat{q} + z_{\alpha/2}^2)q + nm\hat{q}^2 \le 0.$$

The boundaries of the CI are the roots of this quadratic:

$$\frac{2nm\hat{q} + z_{\alpha/2}^2 \pm z_{\alpha/2}\sqrt{1 + 4nm\hat{q}(1 - \hat{q})}}{2(nm + z_{\alpha/2}^2)}.$$

3.15 (a) $\hat{q} = 1001/(40{,}000) = 0.025025$.

(b) There is one degree of freedom. At 5% the critical value is 3.841 (see Table 2.8) and the binomial model is accepted.

Table 2.8 Calculations for Exercise 3.15

Value	Obs.	Exp.	Chi-square
0	9,048	9,035.95	0.016
1	905	927.71	0.556
2+	47	36.34	3.127
Total			3.699

(c) From part (d): $0.025025 \pm 1.96\sqrt{\frac{0.025025(0.974975)}{40{,}000}} = 0.025025 \pm 0.001531$
or $(0.023494, 0.026556)$. From part (e):

$$\frac{80{,}000(0.025025) + 1.96^2 \pm 1.96\sqrt{1 + 160{,}000(0.025025)(0.974975)}}{2(40{,}000 + 1.96^2)}$$

which is $(0.023540, 0.026601)$.

(d) As m increases the *NLL* at the *mle* keeps decreasing. For example, at $m = 5$ it is 3,341.07, at $m = 10$ it is 3,340.03, at $m = 20$ it is 3,339.77. The *mle* is at $m = \infty$ because the sample variance exceeds the sample mean.

(e) The likelihood at $m = \infty$ is 3,339.66 which is the same as for the Poisson. Thus the Poisson is to be preferred.

(f) Because the binomial does not beat the Poisson, use the Poisson as indicated in the solution to Exercise 3.12.

3.16 (a) $\hat{q} = 166/7{,}000 = 0.023714$.

(b) There is one degree of freedom and so the 5% critical value is 3.841 (see Table 2.9) and the binomial model is rejected.

(c)
$$0.023714 \pm 1.96\sqrt{\frac{0.023714(0.976286)}{7{,}000}} = 0.023714 \pm 0.003565$$

$$= (0.020149, 0.027279).$$

Table 2.9 Calculations for Exercise 3.16

Value	Obs.	Exp.	Chi-square
0	861	845.36	0.289
1	121	143.74	3.598
2+	18	10.90	4.625
Total			8.512

The alternative is

$$\frac{14{,}000(0.023714) + 1.96^2 \pm 1.96\sqrt{1 + 28{,}000(0.023714)(0.976286)}}{2(7{,}000 + 1.96^2)}$$

which is $(0.020410, 0.027541)$.

(d) As m increases the NLL at the mle keeps decreasing. For example, at $m = 10$ it is 491.785, at $m = 30$ it is 489.272, at $m = 50$ it is 488.845. The mle is at $m = \infty$ because the sample variance exceeds the sample mean.

(e) The likelihood at $m = \infty$ is 488.241 which is the same as for the Poisson. Thus the Poisson is to be preferred.

(f) Because the binomial does not beat the Poisson, use the geometric as indicated in the solution to Exercise 3.12.

2.5 SECTION 3.5

3.17 For Exercise 3.15 the values at $k = 1, 2, 3$ are 0.1000, 0.0994, and 0.1333 which are nearly constant. The Poisson distribution is recommended. For Exercise 3.16 the values at $k = 1, 2, 3, 4$ are 0.1405, 0.2149, 0.6923, and 1.3333 which is increasing. The geometric/negative binomial is recommended (although the pattern looks more quadratic than linear).

3.18 For the Poisson, $\lambda > 0$ and so it must be $a = 0$ and $b > 0$. For the binomial, m must be a positive integer and $0 < q < 1$. This requires $a < 0$ and $b > 0$ provided $-b/a$ is an integer ≥ 2. For the negative binomial both r and β must be positive so $a > 0$ and b can be anything, provided $b/a > -1$.

The pair $a = -1$ and $b = 1.5$ cannot work because the binomial is the only possibility but $-b/a = 1.5$ which is not an integer. For proof, let p_0 be arbitrary. Then $p_1 = (-1 + 1.5/1)p = 0.5p$ and $p_2 = (-1 + 1.5/2)(0.5p) = -.125p < 0$.

3.19 (a) For $k = 1, 2, 3, 4, 5, 6, 7$ the values are 0.2760, 0.2315, 0.2432, 0.2891, 0.4394, 0.2828, and 0.4268 which are nearly constant. The Poisson model may work well.

(b) The values appear in Table 2.10. Because the sample variance exceeds the sample mean, there is no *mle* for the binomial distribution.

Table 2.10 Calculations for Exercise 3.19

Model	Parameters	NLL	Chi-square	df
Poisson	$\hat{\lambda} = 1.74128$	2,532.86	1,080.80	5
Geometric	$\hat{\beta} = 1.74128$	2,217.71	170.72	7
Negative binomial	$\hat{r} = .867043,\ \hat{\beta} = 2.00830$	2,216.07	165.57	6

(c) The geometric is better than the Poisson by both likelihood and chi-square measures. The negative binomial distribution is not an improvement over the geometric as the *NLL* decreases by only 1.64. When doubled, 3.28 does not exceed the critical value of 3.841. The best choice is geometric, but it does not pass the goodness-of-fit test.

2.6   SECTION 3.6

3.20
$$p_k = p_{k-1} \left[\frac{\beta}{1+\beta} + \frac{r-1}{k} \frac{\beta}{1+\beta} \right] = p_{k-1} \frac{\beta}{1+\beta} \frac{k+r-1}{k}$$
$$= p_{k-2} \left(\frac{\beta}{1+\beta} \right)^2 \frac{k+r-1}{k} \frac{k+r-2}{k-1}$$
$$= p_1 \left(\frac{\beta}{1+\beta} \right)^{k-1} \frac{k+r-1}{k} \frac{k+r-2}{k-1} \cdots \frac{r+1}{2}.$$

The factors will be positive (and thus p_k will be positive) provided $p_1 > 0$, $\beta > 0$, and $r > -1$.

To see that the probabilities sum to a finite amount,
$$\sum_{k=1}^{\infty} p_k = p_1 \sum_{k=1}^{\infty} \left(\frac{\beta}{1+\beta} \right)^{k-1} \frac{k+r-1}{k} \frac{k+r-2}{k-1} \cdots \frac{r+1}{2}$$
$$= p_1 \frac{1}{r} \sum_{k=1}^{\infty} \left(\frac{\beta}{1+\beta} \right)^{k-1} \binom{k+r-1}{k}$$
$$= p_1 \frac{(1+\beta)^{r+1}}{r\beta} \sum_{k=1}^{\infty} \left(\frac{1}{1+\beta} \right)^r \left(\frac{\beta}{1+\beta} \right)^k \binom{k+r-1}{k}.$$

The terms of the summand are the *pf* of the negative binomial distribution and so must sum to a number less than one (p_0 is missing) and so the original sum must converge.

3.21 From the previous solution (with $r = 0$),

$$
\begin{aligned}
1 &= \sum_{k=1}^{\infty} p_k = p_1 \sum_{k=1}^{\infty} \left(\frac{\beta}{1+\beta}\right)^{k-1} \frac{k-1}{k} \frac{k-2}{k-1} \cdots \frac{1}{2} \\
&= p_1 \sum_{k=1}^{\infty} \left(\frac{\beta}{1+\beta}\right)^{k-1} \frac{1}{k} \\
&= p_1 \frac{1+\beta}{\beta} \left[-\log\left(1 - \frac{\beta}{1+\beta}\right) \right]
\end{aligned}
$$

using the Taylor series expansion for $\log(1 - x)$. Thus

$$
p_1 = \frac{\beta}{(1+\beta)\log(1+\beta)}
$$

and

$$
p_k = \left(\frac{\beta}{1+\beta}\right)^k \frac{1}{k\log(1+\beta)}.
$$

3.22

$$
\begin{aligned}
P(z) &= \frac{1}{\log(1+\beta)} \sum_{k=1}^{\infty} \left(\frac{\beta}{1+\beta}\right)^k \frac{1}{k} z^k \\
&= \frac{1}{\log(1+\beta)} \left[-\log\left(1 - \frac{z\beta}{1+\beta}\right) \right] \\
&= \frac{\log\left(\frac{1+\beta}{1+\beta-z\beta}\right)}{\log(1+\beta)} \\
&= 1 - \frac{\log[1 - \beta(z-1)]}{\log(1+\beta)}
\end{aligned}
$$

3.23 The *pgf* goes to $1 - (1 - z)^{-r}$ as $\beta \to \infty$. The derivative with respect to z is $-r(1 - z)^{-r-1}$. The expected value is this derivative evaluated at $z = 1$, which is infinite due to the negative exponent.

3.24 We have

$$
p_k^T = \lambda^k (k!)(e^\lambda - 1)^{-1}
$$

and so

$$
dp_k^T/d\lambda = \frac{k\lambda^{k-1}}{k!(e^\lambda - 1)} - \frac{\lambda^k e^\lambda}{k!(e^\lambda - 1)^2}.
$$

The results for the scoring method are in Table 2.11.

Table 2.11   Calculations for Exercise 3.24

k	n_k	p_k^T	$p_k' = dp_k^T/d\lambda$	$n_k p_k'/p_k^T$	$(p_k')^2/p_k^T$
1	46,545	.91351336	−.47032027	−23,963.587	0.242143
2	3,935	.08142464	.41483539	20,047.706	2.113468
3	317	.00483844	.05179203	3,393.258	0.554397
4	28	.00021563	.00351781	456.795	0.057390
5	3	.00000769	.00016855	65.754	0.003694
6+	0	.00000024	.00000649	0.000	0.000176
Total				0.064	2.971268

The maximum is achieved as the score is essentially zero (0.064). The covariance matrix is diagonal. The diagonal elements are

$$Var(\hat{q}_0) = 0.879337(0.120663)/421,240 = 2.51884 \times 10^{-7}$$
$$Var(\hat{\lambda}) = [50,828(2.971268)]^{-1} = 6.63248 \times 10^{-6}.$$

3.25 We have

$$p_k^T = \frac{1}{k}\left(\frac{\beta}{1+\beta}\right)^k \frac{1}{\log(1+\beta)}$$

and

$$\frac{d}{d\beta}p_k^T = p_k' = \left(\frac{\beta}{1+\beta}\right)^k \frac{1}{(1+\beta)^2\log(1+\beta)}\left[1 - \frac{\beta}{k\log(1+\beta)}\right].$$

Calculations appear in Table 2.12.

Table 2.12   Calculations for Exercise 3.25

k	n_k	p_k^T	$p_k' = dp_k^T/d\lambda$	$n_k p_k'/p_k^T$	$(p_k')^2/p_k^T$
1	46,545	.91822534	−.37499418	−19,008.519	0.153144
2	3,935	.07298279	.29494333	15,902.407	1.191946
3	317	.00773447	.06567286	2,691.625	0.557624
4	28	.00092213	.01193293	362.337	0.154419
5	3	.00011727	.00203934	52.170	0.035464
6+	0	.00001800	.00040572	0.000	0.009145
Total				0.020	2.101742

The maximum is achieved as the score is essentially zero. The covariance matrix is diagonal. The diagonal elements are

$$Var(\hat{q}_0) = 0.879337(0.120663)/421,240 = 2.51884 \times 10^{-7}$$

$$Var(\hat{\beta}) = [50{,}828(2.101742)]^{-1} = 9.36090 \times 10^{-6}.$$

For the goodness-of-fit test, see Table 2.13. There are 3 degrees of freedom, the critical value at 5% significance is 7.815 and so the zero-modified logarithmic distribution is rejected.

Table 2.13   Calculations for Exercise 3.25

Value	Obs.	Exp.	Chi-square
1	46,545	46,671.56	0.343
2	3,935	3,709.57	13.699
3	317	393.13	14.743
4	28	46.87	7.597
5+	3	6.87	2.180
Total			38.562

3.26 For each data set and model, Table 2.14 first gives the negative loglikelihood and then the chi-square test statistic, degrees of freedom, and p-value. If there are not enough degrees of freedom to do the test, no p-value is given.

Table 2.14   Results for Exercise 3.26

	Ex. 3.3	Ex. 3.6	Ex. 3.7	Ex. 3.19
Poisson	3,339.66	488.241	3,578.58	2,532.86
	.00;1;.9773	5.56;1;.0184	16,308;4;0	1,081;5;0
Geometric	3,353.39	477.171	1,132.25	2,217.71
	23.76;2;0	.28;1;.5987	146.84;7;0	170.72;7;0
Neg. bin.	3,339.65	476.457	1,098.64	2,216.07
	.01;0	1.60;0	30.16;6;0	165.57;6;0
ZM Poisson	3,339.66	480.638	2,388.37	2,134.59
	.00;0	1.76;0	900.42;3;0	37.49;6;0
ZM geometric	3,339.67	476.855	1,083.56	2,200.56
	.00;0	1.12;0	1.52;6;.9581	135.43;6;0
ZM logarithmic	3,339.91	475.175	1,171.13	2,308.82
	.03;0	.66;0	186.05;6;0	361.18;6;0
ZM neg. bin.	3,339.63	473.594	1,083.47	2,132.42
	.00;-1	.05;0	1.32;5;.9331	28.43;5;0

For Exercise 3.3, the Poisson is the clear choice. It is the only one parameter distribution acceptable by the goodness-of-fit test and no two parameter distribution improves the NLL by more than 0.03.

For Exercise 3.6, the geometric is the best one parameter distribution and is acceptable by the goodness-of-fit test. The best two paraemter distribution is the negative binomial, but the improvement in the NLL is only .714 which is not significant. (The test statistic with one degree of freedom is 1.428). The three parameter ZM negative binomial improves the NLL by 3.577 over the geometric. This is significant (with two degrees of freedom). So an argument could be made for the ZM negative binomial, but the simpler geometric still looks to be a good choice.

For Exercise 3.7, the best one parameter distribution is the geometric, but it is not acceptable. The best two parameter distribution is the ZM geometric which does pass the goodness-of-fit test and has a much lower NLL. The ZM negative binomial lowers the NLL by 0.09 and is not a significant improvement.

For Exercise 3.19, none of the distributions fit well. According to the NLL, the ZM negative binomial is the best choice, but it does not look very promising.

3.27 The parameter estimates are $\hat{\beta} = 22.6456$ and $\hat{p}_0 = 0.163022$. The estimated variances are 1.63888 and 0.000271264 respectively.
(a) 95% confidence intervals are plus or minus 1.96 standard deviations or 22.6456 ± 2.50917 and 0.163022 ± 0.032281.
(b) Because the estimators are independent the probability is $(0.95)^2 = 0.9216$.
(c) The expected value is $\mu = (1 - p_0^M)(1 + \beta)$ and so the estimated mean is 19.7908. The derivatives are $\partial\mu/\partial\beta = 1 - p_0^M$ and $\partial\mu/\partial p_0^M = -(1 + \beta)$. The estimates are 0.836978 and -23.6456. The estimated variance of $\hat{\mu}$ is $(0.836978)^2(1.63888) + (-23.6456)^2(0.000271264) = 1.29976$ for a standard deviation of 1.14007. The confidence interval is 19.7908 ± 2.2345.

3.28 (a) $p_k/p_{k-1} = \left(\frac{k-1}{k}\right)^{\rho+1} \neq a + b/k$ for any choices of a, b, and ρ, and for all k.
(b) The *mle* is $\hat{\rho} = 3.0416$ using numerical methods.
(c) The test statistic is 785.18 and with 3 degrees of freedom the model is clearly not acceptable.

2.7 SECTION 3.7

3.29 $\bar{x} = 0.155140$ and $s^2 = 0.179314$. $E(N) = \lambda_1\lambda_2$ and $Var(N) = \lambda_1(\lambda_2 + \lambda_2^2)$. Solving the equations yields $\hat{\lambda}_1 = 0.995630$ and $\hat{\lambda}_2 = 0.155821$. For the secondary distribution, $f_j = e^{-0.155821}(0.155821)^j/j!$ and then $g_0 =$

$exp[-0.99563(1 - e^{-0.155821})] = 0.866185$. Then,

$$g_k = \sum_{j=1}^{k} \frac{0.99563}{k} j f_j g_{k-j}.$$

For the goodness-of-fit test with 2 degrees of freedom, see Table 2.15, and the model is clearly rejected.

Table 2.15   Calculations for Exercise 3.29

Value	Obs.	Exp.	Chi-square
0	103,704	103,814.9	0.12
1	14,075	13,782.0	6.23
2	1,766	1,988.6	24.92
3	255	238.9	1.09
4+	53	28.6	20.82
Total			53.18

3.30 The Poisson-Poisson model from Exercise 3.29 has $\lambda_1 = 0.99563$ and $\lambda_2 = 0.155821$. The Poisson-ZT Poisson model has

$$\lambda_1^* = 0.995630(1 - e^{-0.155821}) = 0.143657, \quad \lambda_2^* = 0.155821.$$

The mean of the secondary distribution is

$$0.155821/(1 - e^{-0.155821}) = 1.079933.$$

The mean of the compound distribution is

$$0.143657(1.079933) = 0.155140,$$

matching the earlier first moment. The variance is

$$\lambda_1^* \left\{ \frac{\lambda_2^*[1 - (\lambda_2^* + 1)e^{-\lambda_2^*}]}{(1 - e^{-\lambda_2^*})^2} + \left[\frac{\lambda_2^*}{1 - e^{-\lambda_2^*}} \right]^2 \right\} = 0.179314.$$

$$g_0 = e^{-0.143657} = 0.866185$$
$$g_1 = \frac{0.143657}{1} 0.924112(0.866185) = 0.114991.$$

3.31 Poisson: $P(z) = e^{\lambda(z-1)}$, $B(z) = e^z$, $\theta = \lambda$.

Negative binomial: $P(z) = [1 - \beta(z-1)]^{-r}$, $B(z) = (1-z)^{-r}$, $\theta = \beta$.
Binomial: $P(z) = [1 + q(z-1)]^m$, $B(z) = (1+z)^m$, $\theta = q$.

3.32 Let $\theta = \lambda_1$ and $\beta = \lambda_2$. Then

$$
\begin{aligned}
f_i &= e^{-\beta}\beta^i/i!, \\
g_0 &= \exp(-\theta + \theta e^{-\beta}), \\
g_k &= \sum_{j=1}^{k} \frac{\theta j}{k} f_j g_{k-j}, \\
\frac{\partial g_k}{\partial \theta} &= \sum_{j=1}^{k} \frac{j}{k}\left(f_j g_{k-j} + \theta f_j \frac{\partial g_{k-j}}{\partial \theta} \right), \\
\frac{\partial g_k}{\partial \beta} &= \sum_{j=1}^{k} \frac{\theta j}{k}\left[(f_{j-1} - f_j)g_{k-j} + f_j \frac{\partial g_{k-j}}{\partial \beta} \right].
\end{aligned}
$$

The calculations appear in Table 2.16

The totals of (near) zero in columns (6) and (7) indicate that the maximum has been reached. The covariance matrix is obtained by multiplying the final three column totals by the sample size and then inverting. This yields (using more digits than in the table):

$$
\begin{bmatrix} 13{,}820.89 & 104{,}633.0 \\ 104{,}633.0 & 821{,}647.1 \end{bmatrix}^{-1} = \begin{bmatrix} 0.0020147 & -0.00025656 \\ -0.00025656 & 0.000033889 \end{bmatrix}
$$

Table 2.16 Calculations for Exercise 3.32

(1)	(2)	(3)	(4)	(5)	(6)	(7)	(8)	(9)	(10)
k	n_k	g_k	$\partial g_k/\partial \beta$	$\partial g_k/\partial \theta$	$\frac{(2)(4)}{(3)}$	$\frac{(2)(5)}{(3)}$	$\frac{(4)(4)}{(3)}$	$\frac{(4)(5)}{(3)}$	$\frac{(5)(5)}{(3)}$
0	103,704	0.8653	−0.8277	−0.1137	−99,195	−13,627	0.7917	0.1088	0.0149
1	14,075	0.1166	0.5995	0.0906	72,370	10,932	3.0826	0.4656	0.0703
2	1,766	0.0161	0.1888	0.0196	20,753	2,156	2.2190	0.2305	0.0239
3	255	0.0019	0.0339	0.0031	4,681	425	0.6218	0.0565	0.0051
4	45	0.0002	0.0048	0.0004	1,125	94	0.1193	0.0100	0.0008
5	6	0.0000	0.0006	0.0000	190	15	0.0183	0.0014	0.0001
6	2	0.0000	0.0001	0.0000	77	6	0.0024	0.0002	0.0000
7+	0	0.0000	0.0000	0.0000	0	0	0.0003	0.0000	0.0000
Total	119,853	1.0000	0.0000	0.0000	0.0045	0.0007	6.8555	0.8730	0.1153

3.33 Results appear in Table 2.17. The entries are the negative loglikelihood, the chi-square test statistic, degrees of freedom, and, the p-value (if the degrees of freedom are positive).

Table 2.17   Results for Exercise 3.33

	Ex. 3.3	Ex. 3.6	Ex. 3.7	Ex. 3.19
Poisson-Poisson	3,339.65	478.306	1,198.28	2,151.88
	0.01;0	1.35;0	381.25;6;0	51.85;6;0
Polya-Aeppli	3,339.65	477.322	1,084.95	2,183.48
	0.01;0	1.58;0	4.32;6;0.6335	105.95;6;0
Poisson-I.G.	3,339.65	475.241	1,174.82	2,265.34
	0.01;0	1.30;0	206.08;6;0	262.74;6;0
Poisson-ETNB	3,339.65	473.624	1,083.56	did not
	0.01;−1	0.02;−1	1.52;5;0.9112	converge

For Exercise 3.3, the Poisson cannot be topped. These four improve the loglikelihood by only 0.01 and all have more parameters.

For Exercise 3.6, both the Poisson-inverse Gaussian and Poisson-ETNB improve the loglikelihood over the geometric. The improvements are 1.93 and 3.547. When doubled, they are slightly (p-values of 0.04945 and 0.0288 respectively with 1 and 2 degrees of freedom) significant. The goodness-of-fit test cannot be done. The geometric model, which easily passed the goodness-of-fit test still looks good.

For Exercise 3.7, none of the models improved the loglikelihood over the ZM geometric (although the Poisson-ETNB, with one more parameter, tied). As well, the ZM geometric has the highest p-value and is clearly acceptable.

For Exercise 3.19, none of the models have a superior loglikelihood versus the ZM negative binomial. Although this model is not acceptable, it is the best one from among the available choices.

3.34
$$P_S(z) = \prod_{i=1}^{n} P_{S_i}(z) = \prod_{i=1}^{n} exp\{\lambda_i[P_2(z) - 1]\}$$
$$= exp\{(\sum_{i=1}^{n} \lambda_i)[P_2(z) - 1]\}.$$

This is a compound distribution. The primary distribution is Poisson with parameter $\sum_{i=1}^{n} m\lambda_i$. The secondary distribution has *pgf* $P_2(z)$.

3.35 The geometric-geometric distribution has *pgf*
$$P_{GG}(z) = (1 - \beta_1\{[1 - \beta_2(z - 1)]^{-1} - 1\})^{-1}$$

$$= \frac{1 - \beta_2(z - 1)}{1 - \beta_2(1 + \beta_1)(z - 1)}.$$

The Bernoulli-geometic distribution has *pgf*

$$
\begin{aligned}
P_{BG}(z) &= 1 + q\{[1 - \beta(z - 1)]^{-1} - 1\} \\
&= \frac{1 - \beta(1 - q)(z - 1)}{1 - \beta(z - 1)}.
\end{aligned}
$$

The ZM geometric distribution has *pgf*

$$
\begin{aligned}
P_{ZMG}(z) &= p_0 + (1 - p_0)\frac{[1 - \beta^*(z - 1)]^{-1} - (1 + \beta^*)^{-1}}{1 - (1 + \beta^*)^{-1}} \\
&= \frac{1 - (p_0\beta^* + p_0 - 1)(z - 1)}{1 - \beta^*(z - 1)}.
\end{aligned}
$$

In $P_{BG}(z)$, replace $1 - q$ with $(1 + \beta_1)^{-1}$ and replace β with $\beta_2(1 + \beta_1)$ to see that it matches $P_{GG}(z)$. It is clear that the new parameters will stay within the allowed ranges.

In $P_{GG}(z)$, replace q with $(1 - p_0)(1 + \beta^*)/\beta^*$ and replace β with β^*. Some algebra leads to $P_{ZMG}(z)$.

3.36 The binomial-geometric distribution has *pgf*

$$
\begin{aligned}
P_{BG}(z) &= \left\{1 + q\left[\frac{1}{1 - \beta(z - 1)} - 1\right]\right\}^m \\
&= \left[\frac{1 - \beta(1 - q)(z - 1)}{1 - \beta(z - 1)}\right]^m.
\end{aligned}
$$

The negative binomial-geometric distribution has *pgf*

$$
\begin{aligned}
P_{NBG}(z) &= \left\{1 - \beta_1\left[\frac{1}{1 - \beta_2(z - 1)} - 1\right]\right\}^{-r} \\
&= \left[\frac{1 - \beta_2(1 + \beta_1)(z - 1)}{1 - \beta_2(z - 1)}\right]^{-r} \\
&= \left[\frac{1 - \beta_2(z - 1)}{1 - \beta_2(1 + \beta_1)(z - 1)}\right]^r.
\end{aligned}
$$

In the binomial-geometric *pgf*, replace m with r, $\beta(1 - q)$ with β_2, and β with $(1 + \beta_1)\beta_2$ to obtain $P_{NBG}(z)$.

2.8 SECTION 3.8

3.37 $P(z) = \sum_{k=0}^{\infty} z^k p_k$, $P^{(1)}(z) = \sum_{k=0}^{\infty} k z^{k-1} p_k$,

$$P^{(j)}(z) = \sum_{k=0}^{\infty} k(k - 1) \cdots (k - j + 1) z^{k-j} p_k, \; P^{(j)}(1)$$

$$= \sum_{k=0}^{\infty} k(k-1)\cdots(k-j+1)p_k$$
$$= E[N(N-1)\cdots(N-j+1)].$$

$$P(z) = \exp\{\lambda[P_2(z)-1]\},$$
$$P^{(1)}(z) = \lambda P(z)P_2^{(1)}(z),$$
$$\mu = P^{(1)}(1) = \lambda P(1)m_1' = \lambda m_1'.$$

$$P^{(2)}(z) = \lambda^2 P(z)\left[P_2^{(1)}\right]^2 + \lambda P(z)P_2^{(2)}(z),$$
$$\mu_2' - \mu = P^{(2)}(1) = \lambda^2(m_1')^2 + \lambda(m_2' - m_1').$$

Then $\mu_2' = \lambda^2(m_1')^2 + \lambda m_2'$ and $\mu_2 = \mu_2' - \mu^2 = \lambda m_2'$.

$$P^{(3)}(z) = \lambda^3 P(z)\left[P_2^{(1)}\right]^3 + 3\lambda^2 P(z)P_2^{(1)}(z)P_2^{(2)}(z) + \lambda P(z)P_2^{(3)}(z),$$
$$\mu_3' - 3\mu_2' + 2\mu = \lambda^3(m_1')^3 + 3\lambda^2 m_1'(m_2' - m_1') + \lambda(m_3' - 3m_2' + 2m_1').$$

Then,

$$
\begin{aligned}
\mu_3 &= \mu_3' - 3\mu_2'\mu + 2\mu^3 \\
&= (\mu_3' - 3\mu_2' + 2\mu) + 3\mu_2' - 2\mu - 3\mu_2'\mu + 2\mu^3 \\
&= \lambda^3(m_1')^3 + 3\lambda^2 m_1'(m_2' - m_1') + \lambda(m_3' - 3m_2' + 2m_1') \\
&\quad + 3[\lambda^2(m_1')^2 + \lambda m_2'] - 2\lambda m_1' - 3\lambda m_1'[\lambda^2(m_1')^2 + \lambda m_2'] + 2\lambda^3(m_1')^3 \\
&= \lambda^3[(m_1')^3 - 3(m_1')^3 + 2(m_1')^3] \\
&\quad + \lambda^2[3m_1'm_2' - 3(m_1')^2 + 3(m_1')^2 - 3m_1'm_2'] \\
&\quad + \lambda[m_3' - 3m_2' + 2m_1' + 3m_2' - 2m_1'] \\
&= \lambda m_3'.
\end{aligned}
$$

3.38 For the binomial distribution, $m_1' = mq$ and $m_2' = mq(1-q) + m^2q^2$. For the third central moment, $P^{(3)}(z) = m(m-1)(m-2)[1+q(z-1)]^{m-3}q^3$ and $P^{(3)}(1) = m(m-1)(m-2)q^3$. Then, $m_3' = m(m-1)(m-2)q^3 + 3mq - 3mq^2 + 3m^2q^2 - 2mq$.

$\mu = \lambda m_1' = \lambda mq$.
$\sigma^2 = \lambda m_2' = \lambda(mq - mq^2 + m^2q^2) = \lambda mq(1-q+mq) = \mu[1+q(m-1)]$.

$$\mu_3 = \lambda m_3' = \lambda mq(m^2q^2 - 3mq^2 + 2q^2 + 3 - 3q + 3m - 2)$$

$$
\begin{aligned}
&= \lambda mq(3)(1 - q + mq) - \lambda mq(2) + \lambda mq(m^2q^2 - 3mq^2 + 2q^2) \\
&= 3\sigma^2 - 2\mu + \lambda mq(m^2q^2 - 3mq^2 + 2q^2) \\
&= 3\sigma^2 - 2\mu + \lambda mq(m - 1)(m - 2)q^2.
\end{aligned}
$$

Now, $\frac{m-2}{m-1}\frac{(\sigma^2-\mu)^2}{\mu} = \frac{m-2}{m-1}\frac{\mu^2[q(m-1)]^2}{\mu} = (m-2)(m-1)\lambda mq^3$ and the relationship is shown.

3.39 The coefficient discussed in the section is $\frac{(\mu_3 - 3\sigma^2 + 2\mu)\mu}{(\sigma^2 - \mu)^2}$. For the five data sets, use the empirical estimates. For the last two sets, the final category is for some number or more. For these cases, estimate by assuming the observations were all at that highest value. The five coefficient estimates are (a) -0.85689, (b) $-1,817.27$, (c) -5.47728, (d) 1.48726, (e) -0.42125. For all but data set (d) it appears that a compound Poisson model will not be appropriate. For data set (d) it appears that Polya-Aeppli model will do well. Table 2.18 summarizes a variety of fits. The entries are the negative loglikelihood, the chi-square test statistic, the degrees of freedom, and the p-value.

Table 2.18 Results for Exercise 3.39

	(a)	(b)	(c)	(d)	(e)
Poiss.	36,188.3	206.107	481.886	1,461.49	841.113
	190.75;2;0	0.06;1;.8021	2.40;2;.3017	267.52;2;0	6.88;6;.3323
Geom.	36,123.6	213.052	494.524	1,309.64	937.975
	41.99;3;0	14.02;2;0	24.03;3;0	80.13;4;0	189.47;9;0
NB	36,104.1	did not	481.054	1,278.59	837.455
	0.09,1,.7631	converge	0.31;1;.5779	1.57;4;.8139	3.46;6;.7963
P-bin.	36,106.9	did not	481.034	1,320.71	837.438
$m = 2$	2.38;1;.1228	converge	0.38;1;.5357	69.74;2;0	3.10;6;.7963
P-bin.	36,106.1	did not	481.034	1,303.00	837.442
$m = 3$	1.20;1;.2728	converge	0.35;1;.5515	63.10;3;0	3.19;6;.7853
Polya-	36,104.6	did not	481.044	1,280.59	837.452
Aeppli	0.15;1;.6966	converge	0.32;1;.5731	6.52;4;.1638	3.37;6;.7615
Ney.-A	36,105.3	did not	481.037	1,288.56	837.448
	0.49;1;.4820	converge	0.33;1;.5642	24.14;3;0	3.28;6;.7736
P-iG	36,103.6	did not	481.079	1,281.90	837.455
	0.57;1;.4487	converge	0.31;1;.5761	8.75;4;.0676	3.63;6;.7260
P-ETNB	36,103.5	did not	did not	1,278.58	did not
	5.43;1;.0198	converge	converge	1.54;3;.6740	converge

(a) All two-parameter distributions are superior to the two one-parameter distributions. The best two-parameter distributions are negative binomial (best loglikelihood) and Poisson-inverse Gaussian (best p-value). The three-parameter Poisson-ETNB is not a significant improvement by the likelihood ratio test. The simpler negative binomial is an excellent choice.

(b) Because the sample variance is less than the sample mean, mle's exist only for the Poisson and geometric distributions. The Poisson is clearly acceptable by the goodness-of-fit test.

(c) None of the two-parameter models are significant imiprovements over the Poisson according to the likelihood ratio test. Even though some have superior p-values, the Poisson is acceptable and should be our choice.

(d) The two-parameter models are better by the likelihood ratio test. The best is negative binomial on all measures. The Poisson-ETNB is not better and so use the negative binomial, which passes the goodness-of-fit test. The moment analysis supported the Polya-Aeppli, which was acceptable, but not as good as the negative binomial.

(e) The two-parameter models are better by the likelihood ratio test. The best is Poisson-binomial with $m = 2$, though the simpler and more popular negative binomial is a close alternative.

2.9   SECTION 3.9

3.40 $P(z) = \int_0^\infty z^x \frac{\mu}{(2\pi\beta x^3)^{1/2}} \exp[-(x-\mu)^2/(2\beta x)]dx.$

The integral can be evaluated by replacing z^x with $\exp(\log z)$, making the notational substituion $\theta = 1 - 2\beta \log z$, and then completing the square as noted below.

$$
\begin{aligned}
P(z) &= \frac{\mu}{(2\pi\beta)^{1/2}} \int_0^\infty \frac{1}{x^{3/2}} \exp\left(-\frac{\theta x}{2\beta} + \frac{\mu}{\beta} - \frac{\mu^2}{2\beta x}\right) dx \\
&= \frac{\mu}{(2\pi\beta)^{1/2}} \int_0^\infty \frac{1}{x^{3/2}} \exp\left[-\frac{\theta}{2\beta}\left(x - \frac{2\mu}{\theta} + \frac{\mu^2}{\theta x}\right)\right] dx \\
&= \frac{\mu}{(2\pi\beta)^{1/2}} \int_0^\infty \frac{1}{x^{3/2}} \exp\left[-\frac{\theta}{2\beta}\left(x^{1/2} - \frac{\mu}{\theta^{1/2}x^{1/2}}\right)^2 \right. \\
&\qquad \left. + \frac{\mu}{\beta} - \frac{\mu\theta^{1/2}}{\beta}\right] dx \\
&= \frac{\mu}{(2\pi\beta)^{1/2}} \exp\left[\frac{\mu}{\beta}\left(1 - \theta^{1/2}\right)\right] \\
&\qquad \times \int_0^\infty \frac{1}{x^{3/2}} \exp\left[-\frac{\theta}{2\beta x}\left(x - \frac{\mu}{\theta^{1/2}}\right)^2\right] dx.
\end{aligned}
$$

The integral is that for an inverse Gaussian distribution with 'μ'$= \mu/\theta^{1/2}$ and 'β'$= \beta/\theta$ and so the integral is $(2\pi\beta/\theta)^{1/2}/(\mu/\theta^{1/2})$ and thus

$$P(z) = \exp\left[\frac{\mu}{\beta}\left(1 - \theta^{1/2}\right)\right] = \exp\left\{\frac{\mu}{\beta}\left[1 - (1 - 2\beta\log z)^{1/2}\right]\right\}.$$

3.41
$$
\begin{aligned}
P(z) &= \int_0^\infty z^x \frac{e^\lambda}{\mu\lambda^{1/\alpha}} e^{-x/\mu} f_\alpha\left(\frac{x}{\mu\lambda^{1/\alpha}}\right) dx \\
&= e^\lambda \int_0^\infty z^{\mu\lambda^{1/\alpha}y} e^{-y\lambda^{1/\alpha}} f_\alpha(y) dy \\
&= e^\lambda \int_0^\infty \left(z^{\mu\lambda^{1/\alpha}} e^{-\lambda^{1/\alpha}}\right)^y f_\alpha(y) dy \\
&= e^\lambda P_\alpha\left(z^{\mu\lambda^{1/\alpha}} e^{-\lambda^{1/\alpha}}\right) \\
&= e^\lambda \exp[-(\mu\lambda^{1/\alpha}\log z + \lambda^{1/\alpha})^\alpha] \\
&= \exp\{\lambda[1 - (1 - \mu\log z)^\alpha]\}.
\end{aligned}
$$

3.42 For the compound distribution, the *pgf* is $P(z) = P_{NB}[P_P(z)]$ whch is exactly the same as for the mixed distribution because mixing distribution goes on the outside.

3.43
$$P_N(z) = \prod_{i=1}^n P_{N_i}(z) = \prod_{i=1}^n P_i[e^{\lambda(z-1)}]$$

Then N is mixed Poisson. The mixing distribution has *pgf* $P(z) = \prod_{i=1}^n P_i(z)$.

3.44 Because Θ has a scale parameter, we can write $\Theta = cX$ where X has density (probability, if discrete) function $f_X(x)$. Then, for the mixed distribution (with formulas for the continuous case)

$$
\begin{aligned}
p_k &= \int \frac{e^{-\lambda\theta}(\lambda\theta)^k}{k!} f_\Theta(\theta) d\theta \\
&= \frac{\lambda^k}{k!} \int e^{-\lambda\theta}\theta^k f_X(\theta/c)c^{-1} d\theta \\
&= \frac{\lambda^k}{k!} \int e^{-\lambda cx}c^k x^k f_X(x) dx.
\end{aligned}
$$

The parameter λ appears only as the product λc. Therefore, the mixed distribution does not depend on λ as a change in c will lead to the same distribution.

3.45 $p_k = \displaystyle\int_0^\infty \frac{e^{-\theta}\theta^k}{k!} \frac{\alpha^2}{\alpha + 1}(\theta + 1)e^{-\alpha\theta} d\theta$

$$= \frac{\alpha^2}{k!(\alpha+1)} \int_0^\infty \theta^k(\theta+1)e^{-(\alpha+1)\theta}\,d\theta$$

$$= \frac{\alpha^2}{k!(\alpha+1)} \int_0^\infty \beta^k(\alpha+1)^{-k-1}(\beta/(\alpha+1)+1)e^{-\beta}\,d\beta$$

$$= \frac{\alpha^2}{k!(\alpha+1)^{k+2}}[\Gamma(k+2)/(\alpha+1)+\Gamma(k+1)]$$

$$= \frac{\alpha^2[(k+1)/(\alpha+1)+1]}{(\alpha+1)^{k+2}}.$$

The *pgf* of the mixing distribution is

$$P(z) = \int_0^\infty z^\theta \frac{\alpha^2}{(\alpha+1)}(\theta+1)e^{-\alpha\theta}\,d\theta$$

$$= \frac{\alpha^2}{\alpha+1} \int_0^\infty (\theta+1)e^{-(\alpha-\log z)\theta}\,d\theta$$

$$= \frac{\alpha^2}{\alpha+1}\left[-\frac{\theta+1}{\alpha-\log z}e^{-(\alpha-\log z)\theta} - \frac{\theta+1}{(\alpha-\log z)^2}e^{-(\alpha-\log z)\theta} \right]\Big|_0^\infty$$

$$= \frac{\alpha^2}{\alpha+1}\left[\frac{1}{\alpha-\log z} + \frac{1}{(\alpha-\log z)^2} \right]$$

and so the *pgf* of the mixed distribution (obtained by replacing $\log z$ with $\lambda(z-1)$) is

$$P(z) = \frac{\alpha^2}{\alpha+1}\left\{ \frac{1}{\alpha-\lambda(z-1)} + \frac{1}{[\alpha-\lambda(z-1)]^2} \right\}$$

$$= \frac{\alpha}{\alpha+1}\frac{\alpha}{\alpha-\lambda(z-1)} + \frac{1}{\alpha+1}\left[\frac{\alpha}{\alpha-\lambda(z-1)} \right]^2$$

which is a two-point mixture of negative binomial variables with identical values of β. Each is also a Poisson-logarithmic distribution with identical logarithmic secondary distributions. The equivalent distribution has a logarithmic secondary distribution with $\beta = \lambda/\alpha$ and a primary distribution that is a mixture of Poissons. The first Poisson has $\lambda = \log(1+\lambda/\alpha)$ and the second Poisson has $\lambda = 2\log(1+\lambda/\alpha)$. To see that this is correct, note that the *pgf* is

$$P(z) = \frac{\alpha}{\alpha+1}\exp\left\{ \log(1+\lambda/\alpha)\frac{\log[1-\lambda(z-1)/\alpha]}{\log(1+\lambda/\alpha)} \right\}$$

$$\qquad + \frac{1}{\alpha+1}\exp\left\{ 2\log(1+\lambda/\alpha)\frac{\log[1-\lambda(z-1)/\alpha]}{\log(1+\lambda/\alpha)} \right\}$$

$$= \frac{\alpha}{\alpha+1}\log[1-\lambda(z-1)/\alpha] + \frac{1}{\alpha+1}[1-\lambda(z-1)/\alpha]^2.$$

3.46 EXCEL solver reports the following *mle*'s to four decimals: $\hat{p} = 0.9312$, $\hat{\lambda}_1 = 0.1064$, and $\hat{\lambda}_2 = 0.6560$. The negative loglikelihood is 10,221.9. Rounding these numbers to two decimals produces a negative loglikelihod of 10,223.3 while Tröbliger's solution is superior at 10,222.1. A better two decimal solution is (0.94,0.11,0.69) which gives 10,222.0. The negative binomial distribution was found to have a negative loglihood of 10,223.4. The extra parameter for the two-point mixture cannot be justified (using the likelihood ratio test).

2.10   SECTION 3.10

3.47
$$L = \prod_{k=1}^{6} \frac{(n_k + e_k + 1)\cdots e_k}{n_k!} \left(\frac{1}{1+\beta}\right)^{e_k} \left(\frac{\beta}{1+\beta}\right)^{n_k}$$
$$\propto \beta^{\Sigma n_k}(1+\beta)^{-\Sigma(n_k+e_k)}.$$

The logarithm is

$$(\log\beta)\sum_{k=1}^{6} n_k - [\log(1+\beta)]\sum_{k=1}^{6}(n_k+e_k)$$

and setting the derivative equal to zero yields

$$\beta^{-1}\sum_{k=1}^{6} n_k - (1+\beta)^{-1}\sum_{k=1}^{6}(n_k+e_k) = 0$$

for $\hat{\beta} = \sum_{k=1}^{6} n_k / \sum_{k=1}^{6} e_k = 0.09772$. The expected number is $E_k = \hat{\beta}e_k$ which is exactly the same as for the Poisson model. The goodness-of-fit test produces the same results, so the models are impossible to distinguish.

3.48 (a) For frequency, the probability of a claim is 0.03 and for severity, the probability of a 10,000 claim is 1/3 and of a 20,000 claim is 2/3.
(b) $\Pr(X = 0) = 1/3$ and $\Pr(X = 10,000) = 2/3$.
(c) For frequency, the probability of a claim is 0.02 and the severity distribution places probability one at 10,000.

3.49 $\Pr(X > 500) = \left(\frac{1000}{1000+500}\right)^2 = 4/9$. The original frequency distribution is P-ETNB with $\lambda = 3$, $r = -0.5$, and $\beta = 2$. The new frequency is P-ZMNB with $\lambda = 3$, $r = -0.5$, $\beta = 2(4/9) = 8/9$ and $p_0^T = \frac{0 - 3^{.5} + (17/9)^{.5} - 0(17/9)^{.5}}{1-3^{.5}} = 0.4886$. This is equivalent to a P-ETNB with $\lambda = 3(1 - 0.4886) = 1.53419$, $\beta = 8/9$, and $r = -0.5$. which is P-IG with $\lambda = 1.53419$ and $\beta = 8/9$.
   The new severity distribution has *cdf* $F(x) = \frac{F(x+500)=F(500)}{1-F(500)} = 1 - \left(\frac{1,500}{1,500+x}\right)^2$ which is Pareto with $\alpha = 2$ and $\theta = 1,500$.

3.50 The value of p_0^M will be negative and so there is no appropriate frequency distribution to describe effect of lowering the deductible.

3.51 There were 7 years with 0 hurricanes, 18 years with 1, 5 with 2, 1 with 3 and 1 with 4. The results of fitting a number of distributions appear in Table 2.19 (no others produced *mle*'s):

Table 2.19 Calculations for Exercise 3.51

Model	NLL	Chi-square	df	p-value
Poisson	40.2991	5.33	1	0.0209
Geometric	46.3737	17.40	1	0.0000
ZM Poisson	38.0683	0.15	0	
ZM geometric	37.7497	0.00	0	
Negative binomial	37.8285	0.05	0	

By the likelihood ratio test, the two parameter models are significant improvements over the corresponding one-parameter models. The best two parameter model, by both *NLL* and chi-square is the ZM geometric. The parameter estimates are $\hat{p}_0 = 0.21875$ and $\hat{\beta} = 0.4$. The best choice for the ground-up severity model is inverse Gaussian with $\mu = 204{,}500$ and $\theta = 38{,}060.3$. For hurricanes with damage in excess of 500,000 (thousand) the frequency distribution remains ZM geometric, but the parameters are $\hat{p}_0 = 0.897595$ and $\hat{\beta} = 0.0389082$.

3

Chapter 4 Solutions

3.1 SECTION 4.1

4.1 $(c) + (d) + (e)$ is:

For $X \leq 1{,}000$,	$0 + 0 + X = X$
For $1{,}000 < X \leq 63{,}500$,	$0.8(X - 1{,}000) + 800 + 0.2X = X$
For $63{,}500 < X \leq 126{,}000$,	$0.8(X - 63{,}500) + 50{,}000 + 800 + 0.2X = X$
For $X > 126{,}000$,	$50{,}000 + 50{,}000 + X - 100{,}000 = X.$

3.2 SECTION 4.2

4.2
$$
\begin{aligned}
E(N) &= P^{(1)}(1), \\
P^{(1)}(z) &= \alpha Q(z)^{\alpha-1} Q^{(1)}(z), \\
P^{(1)}(1) &= \alpha Q(1)^{\alpha-1} Q^{(1)}(1) \propto \alpha.
\end{aligned}
$$

4.3 The Poisson and all compound distributions with a Poisson primary distribution have a *pgf* of the form $P(z) = \exp\{\lambda[P_2(z) - 1]\} = [Q(z)]^\lambda$ where $Q(z) = \exp[P_2(z) - 1]$.

The negative binomial and geometric distributions and all compound distributions with a negative binomial or geometric primary distribution have $P(z) = \{1 - \beta[P_2(z) - 1]\}^{-r} = [Q(z)]^r$ where $Q(z) = \{1 - \beta[P_2(z) - 1]\}^{-1}$.

The same is true for the binomial distribution and binomial-X compound distributions with $\alpha = m$ and $Q(z) = 1 + q[P_2(z) - 1]$.

The zero-truncated and zero-modified distributions cannot be written in this form.

3.3   SECTION 4.3

4.4 To simplify writing the expressions, let

$$
\begin{aligned}
Njp &= \mu'_{Nj} = E(N^j) \\
Nj &= \mu_{Nj} = E[(N - N1p)^j] \\
Xjp &= \mu'_{Xj} = E(X^j) \\
Xj &= \mu_{Xj} = E[(XN - X1p)^j]
\end{aligned}
$$

and similarly for S. For the first moment, $P_S^{(1)}(z) = P_N^{(1)}[P_X(z)]P_X^{(1)}(z)$ and so

$$
\begin{aligned}
E(S) &= P_S^{(1)}(1) = P_N^{(1)}[P_X(1)]P_X^{(1)}(1) \\
&= P_N^{(1)}(1)P_X^{(1)}(1) = (N1p)(X1p) = E(N)E(X).
\end{aligned}
$$

For the second moment use

$$
\begin{aligned}
P_S^{(2)}(1) &= S2p - S1p = P_N^{(2)}[P_X(z)][P_X^{(1)}(z)]^2 + P_N^{(1)}[P_X(z)]P_X^{(2)}(z) \\
&= (N2p - N1p)(Xp1)^2 + N1p(X2p - X1p).
\end{aligned}
$$

$$
\begin{aligned}
Var(S) &= S2 = S2p - (S1p)^2 = S2p - S1p + S1p - (S1p)^2 \\
&= (N2p - N1p)(X1p)^2 + N1p(X2p - X1p) + (N1p)(X1p) \\
&\quad -(N1p)^2(X1p)^2 \\
&= N1p[X2p - (X1p)^2] + [N2p - (N1p)^2](X1p)^2 \\
&= (N1p)(X2) + (N2)(X1p)^2.
\end{aligned}
$$

For the third moment use

$$
\begin{aligned}
P_S^{(3)} &= S3p - 3S2p + 2S1p = P_N^{(3)}[P_X(z)][P_X^{(1)}(z)]^3 \\
&\quad +3P_N^{(2)}[P_X(z)]P_X^{(1)}(z)P_X^{(2)}(z) + P_N^{(1)}[P_X(z)]P_X^{(3)}(z) \\
&= (N3p - 3N2p + 2N1p)(X1p)^3 \\
&\quad +3(N2p - N1p)(X1p)(X2p - X1p) \\
&\quad +N1p(X3p - 3X2p + 2X1p).
\end{aligned}
$$

$$
S3 = S3p - 3(S2p)(S1p) + 2(S1p)^3
$$

$$
\begin{aligned}
&= S3p - 3S2p + 2S1p + 3[S2 + (S1p)^2](1 - 3S1p) + 2(S1p)^3 \\
&= (N3p - 3N2p + 2N1p)(X1p)^3 \\
&\quad +3(N2p - N1p)(X1p)(X2p - X1p) \\
&\quad +N1p(X3p - 3X2p + 2X1p) + 3\{N1p[X2p - (X1p)^2] \\
&\quad\quad +[N2p - (N1p)^2](X1p)^2 \\
&\quad\quad +(N1p)^2(X1p)^2\}[1 - 3(N1p)(X1p)] \\
&\quad +2(N1p)^3(X1p)^3 \\
&= N1p(X3) + 3(N2)(X1p)(X2) + N3(X1p)^3.
\end{aligned}
$$

4.5

$$
\begin{aligned}
E(X) &= 1{,}000 + 0.8(500) = 1{,}400. \\
Var(X) &= Var(X_1) + Var(X_2) + 2Cov(X_1, X_2) \\
&= 500^2 + 0.64(300)^2 + 2(0.8)(100{,}000) \\
&= 467{,}600. \\
E(S) &= E(N)E(X) = 4(1{,}400) = 5{,}600, \\
Var(S) &= E(N)Var(X) + Var(N)E(X)^2 \\
&= 4(467{,}600) + 4(1{,}400)^2 = 9{,}710{,}400.
\end{aligned}
$$

4.6 $E(S) = 15(5)(5) = 375.$
$Var(S) = 15(5)(100/12) + 15(5)(6)(5)^2 = 11{,}875.$ $StDev(S) = 108.97.$
The 95th percentile is $375 + 1.645(108.97) = 554.26.$

4.7
$$
\begin{aligned}
Var(N) &= E[Var(N|\lambda)] + Var[E(N|\lambda)] = E(\lambda) + Var(\lambda), \\
E(\lambda) &= 0.25(5) + 0.25(3) + 0.5(2) = 3, \\
Var(\lambda) &= 0.25(25) + 0.25(9) + 0.5(4) - 9 = 1.5, \\
Var(N) &= 4.5.
\end{aligned}
$$

4.8 The calculations appear in Table 3.1.

4.9 The calculations appear in Table 3.2.
$0.06 = f_S(5) = 0.1 - 0.05p,\ p = 0.8.$

4.10 If all 10 do not have AIDS,

$$
\begin{aligned}
E(S) &= 10(1{,}000) = 10{,}000 \\
Var(S) &= 10(250{,}000) = 2{,}500{,}000
\end{aligned}
$$

Table 3.1 Results for Exercise 4.8

x	$f_1(x)$	$f_2(x)$	$f_{1+2}(x)$	$f_3(x)$	$f_S(x)$
0	0.9	0.5	0.45	0.25	0.1125
1	0.1	0.3	0.32	0.25	0.1925
2		0.2	0.21	0.25	0.2450
3			0.02	0.25	0.2500
4					0.1375
5					0.0575
6					0.0050

Table 3.2 Results for Exercise 4.9

x	$f_2(x)$	$f_3(x)$	$f_{2+3}(x)$	$f_1(x)$	$f_S(x)$
0	0.6	0.25	0.150	p	$0.15p$
1	0.2	0.25	0.200	$1-p$	$0.15 + 0.05p$
2	0.1	0.25	0.225		$0.2 + 0.025p$
3	0.1	0.25	0.250		$0.225 + 0.025p$
4			0.100		$0.25 - 0.15p$
5			0.050		$0.1 - 0.05p$
6			0.025		$0.05 - .025p$
7					$0.025 - 0.025p$

and so the premium is $10,000 + 0.1(2,500,000)^{1/2} = 10,158$.

If the number with AIDS has the binomial distribution with $m = 10$ and $q = 0.01$, then, letting N be the number with AIDS,

$$
\begin{aligned}
E(S) &= E[E(S|N)] = E[70,000N + 1,000(10 - N)] \\
&= 10,000 + 69,000[10(0.01)] \\
&= 16,900 \\
Var(S) &= Var[E(S|N)] + E[Var(S|N)] \\
&= Var[70,000N + 1,000(10 - N)] \\
&\quad + E[1,600,000N + 250,000(10 - N)] \\
&= 69,000^2[10(0.01)(0.99)] + 2,500,000 + 1,350,000[10(0.01)] \\
&= 473,974,000
\end{aligned}
$$

and so the premium is $16,900 + 0.1(473,974,000)^{1/2} = 19,077$. The ratio is $10,158/19,077 = 0.532$.

4.11 Let M be the random number of males and C be the number of cigarettes smoked. Then $E(C) = E[E(C|M)] = E[6M + 3(8 - M)] = 3E(M) + 24$. But M has the binomial distribution with mean $8(0.4) = 3.2$ and so $E(C) = 3(3.2) + 24 = 33.6$.

$$
\begin{aligned}
Var(C) &= E[Var(C|M)] + Var[E(C|M)] \\
&= E[64M + 31(8 - M)] + Var(3M + 24) \\
&= 33E(M) + 248 + 9Var(M) \\
&= 33(3.2) + 9(8)(0.4)(0.6) = 370.88.
\end{aligned}
$$

4.12 For insurer A, the group pays the net premium, so the expected total cost is just the expected total claims, that is, $E(S) = 5$.

For insurer B, the cost to the group is $7 - D$ where D is the dividend. We have

$$
D = \begin{cases} 7k - S, & S < 7k \\ 0, & S \geq 7k. \end{cases}
$$

Then $E(D) = \int_0^{7k}(7k - s)(0.1)ds = 2.45k^2$. We want $5 = E(7 - D) = 7 - 2.45k^2$, and so $k = 0.9035$.

4.13 Let θ be the underwriter's estimated mean. The underwriter computes the premium as

$$
\begin{aligned}
2E[(S - 1.25\theta)^+] &= 2\int_{1.25\theta}^{\infty}(s - 1.25\theta)\theta^{-1}e^{-s/\theta}ds \\
&= -2(s - 1.25\theta)e^{-s/\theta} - 2\theta e^{-s/\theta}\Big|_{1.25\theta}^{\infty} \\
&= 2\theta e^{-1.25}.
\end{aligned}
$$

Let μ be the true mean. Then $\theta = .9\mu$. The true expected loss is

$$
E[(S - 1.25(.9)\mu)^+] = 2\int_{1.125\mu}^{\infty}(s - 1.125\mu)\mu^{-1}e^{-s/\mu}ds = \mu e^{-1.125}
$$

The loading is

$$
\frac{2(0.9)\mu e^{-1.25}}{\mu e^{-1.125}} - 1 = 0.5885.
$$

4.14 A convenient relationship is, for discrete distributions on whole numbers, $E[(S - d - 1)^+] = E[(S - d)^+] - 1 + F(d)$. For this problem, $E[(S - 0)^+] = E(S) = 4$. $E[(S - 1)^+] = 4 - 1 + 0.05 = 3.05$, $E[(S - 2)^+] = 3.05 - 1 + 0.11 = 2.16$, $E[(S - 3)^+] = 2.16 - 1 + 0.36 = 1.52$. Then, by linear interpolation, $d = 2 + (2 - 2.16)/(1.52 - 2.16) = 2.25$.

4.15 $15 = \int_{100}^{\infty}[1 - F(s)]ds$, $10 = \int_{120}^{\infty}[1 - F(s)]ds$, $F(120) - F(80) = 0$. Subtracting the second equality from the first yields $5 = \int_{100}^{120}[1 - F(s)]ds$, but over this range $F(s) = F(80)$ and so

$$5 = \int_{100}^{120}[1 - F(s)]ds = \int_{100}^{120}[1 - F(80)]ds = 20[1 - F(80)]$$

and therefore $F(80) = 0.75$.

4.16

$$E(A) = \int_{50k}^{100}\left(\frac{x}{k} - 50\right)(0.01)dx = \left(\frac{x^2}{2k} - 50x\right)(0.01)\Big|_{50k}^{100}$$

$$= 50k^{-1} - 62.5 + 25k.$$

$$E(B) = \int_{0}^{100}kx(0.01)dx = \frac{kx^2}{2}(0.01)\Big|_{50k}^{100} = 50k.$$

The solution to $50k^{-1} - 62.5 + 25k = 50k$ is $k = 0.63746$.

4.17 $E(X) = 440$, $F(30) = 0.3$, $f(x) = 0.01$ for $0 < x \le 30$.

$$\begin{aligned}
E(\text{benefits}) &= \int_{0}^{30} 20x(0.01)dx + \int_{30}^{\infty}[600 + 100(x - 30)]f(x)dx \\
&= 90 + \int_{30}^{\infty}(-2,400)f(x)dx + 100\int_{30}^{\infty} xf(x)dx \\
&= 90 - 2,400[1 - F(30)] + 100\int_{0}^{\infty} xf(x)dx \\
&\quad -100\int_{0}^{30} xf(x)dx \\
&= 90 - 2,400(0.7) + 100(440) - 100\int_{0}^{30} x(0.01)dx \\
&= 90 - 1,680 + 44,000 - 450 = 41,960.
\end{aligned}$$

4.18

$$\begin{aligned}
E(S) &= E(N)E(X) = [0(0.5) + 1(0.4) + 3(0.1)][1(0.9) + 10(0.1)] \\
&= 0.7(1.9) = 1.33.
\end{aligned}$$

We require $\Pr(S > 3.99)$. Using the calculation in Table 3.3, $\Pr(S > 3.99) = 1 - 0.5 - 0.36 - 0.0729 = 0.0671$.

4.19 For 100 independent lives, $E(S) = 100mq$ and $Var(S) = 100m^2q(1 - q) = 250,000$. The premium is $100mq + 500$. For this particular group,

$$E(S) = 97(mq) + (3m)q = 100mq$$

Table 3.3 Calculations for Exercise 4.18

x	$f_X^{*0}(x)$	$f_X^{*1}(x)$	$f_X^{*2}(x)$	$f_X^{*3}(x)$	$f_S(x)$
0	1	0	0	0	0.5000
1	0	0.9	0	0	0.3600
2	0	0	0.81	0	0
3	0	0	0	0.729	0.0729
p_n	0.5	0.4	0	0.1	

$$
\begin{aligned}
Var(S) &= 97m^2q(1-q) + (3m)^2q(1-q) \\
&= 103m^2q(1-q) = 257{,}500
\end{aligned}
$$

and the premium is $100mq + 507.44$. The difference is 7.44.

4.20
$$
\begin{aligned}
E(S) &= 1(8{,}000)(0.025) + 2(8{,}000)(0.025) = 600 \\
Var(S) &= 1(8{,}000)(0.025)(0.975) \\
&\quad + 2^2(8{,}000)(0.025)(0.975) = 975.
\end{aligned}
$$

The cost of reinsurance is $0.03(2)(4{,}500) = 270$.

$$
\begin{aligned}
\Pr(S + 270 > 1{,}000) &= \Pr(S > 730) \\
&= \Pr\left(\frac{S - 600}{\sqrt{975}} > \frac{730 - 600}{\sqrt{975}} = 4.163\right)
\end{aligned}
$$

so $K = 4.163$.

4.21
$$
\begin{aligned}
E(Z) &= \int_{10}^{100} 0.8(y - 10)(0.02)(1 - 0.01y)\,dy \\
&= 0.016 \int_{10}^{100} -0.01y^2 + 1.1y - 10\,dy = 19.44.
\end{aligned}
$$

4.22
$$
\begin{aligned}
\Pr(S > 100) &= \sum_{n=0}^{3} \Pr(N = n)\Pr(X^{*n} > 100) \\
&= 0.5(0) + 0.2\Pr(X > 100) \\
&\quad + 0.2\Pr(X^{*2} > 100) + 0.1\Pr(X^{*3} > 100).
\end{aligned}
$$

Because $X \sim N(100, 9)$, $X^{*2} \sim N(200, 18)$, and $X^{*3} \sim N(300, 27)$ and so

$$
\begin{aligned}
\Pr(S > 100) &= 0.2\Pr\left(Z > \frac{100 - 100}{3}\right) + 0.2\Pr\left(Z > \frac{100 - 200}{\sqrt{18}}\right) \\
&\quad + 0.2\Pr\left(Z > \frac{100 - 300}{3\sqrt{27}}\right) \\
&= 0.2(0.5) + 0.2(1) + 0.1(1) = 0.4.
\end{aligned}
$$

4.23 The calculations are in Table 3.4. The expected retained payment is $2,000(0.1) + 3,000(0.15) + 4,000(0.06) + 5,000(0.6275) = 4,027.5$ and the total cost is $4,027.5 + 1,472 = 5,499.5$.

Table 3.4   Calculations for Exercise 4.23

x	$f_X^{*0}(x)$	$f_X^{*1}(x)$	$f_X^{*2}(x)$	$f_S(x)$
0	1	0	0	0.0625
1	0	0	0	0.0000
2	0	0.4	0	0 .1000
3	0	0.6	0	0.1500
4	0	0	0.16	0.0600
p_k	1/16	1/4	3/8	

4.24 In general, paying for days a through b, the expected number of days is

$$\sum_{k=a}^{b}(k - a + 1)p_k + (b - a + 1)[1 - F(b)]$$

$$= \sum_{k=a}^{b}\sum_{j=a}^{k}p_k + (b - a + 1)[1 - F(b)]$$

$$= \sum_{j=a}^{b}\sum_{k=j}^{b}p_k + (b - a + 1)[1 - F(b)]$$

$$= \sum_{j=a}^{b}[F(b) - F(j - 1)] + (b - a + 1)[1 - F(b)]$$

$$= \sum_{j=a}^{b}[1 - F(j - 1)] = \sum_{j=a}^{b}(0.8)^{j-1}$$

$$= \frac{(0.8)^{a-1} - (0.8)^{b}}{.2}.$$

For the four through ten policy the expected numbered of days is $(0.8^3 - 0.8^{10})/0.2 = 2.02313$. For the four through seventeen policy the expected number of days is $(0.8^3 - 0.8^{17})/0.2 = 2.44741$. This is a 21% increase.

4.25

$$E(S) = \int_1^{\infty} x3x^{-4}dx = 3/2$$

$$E(X^2) = \int_1^{\infty} x^2 3x^{-4}dx = 3$$

$$Var(S) = 3 - (3/2)^2 = 3/4.$$

$$0.9 = \Pr[S \le (1 + \theta)(3/2)] = \int_{1}^{(1+\theta)(3/2)} 3x^{-4}dx$$

$$= 1 - [(1 + \theta)(3/2)]^{-3},$$

and so $\theta = 0.43629$.

$$0.9 = \Pr[S \le 1 + \lambda\sqrt{3/4}] = \int_{1}^{1+\lambda\sqrt{3/4}} 3x^{-4}dx$$

$$= 1 - [1 + \lambda\sqrt{3/4}]^{-3},$$

and so $\lambda = 0.75568$.

3.4   SECTION 4.4

4.26 (a) The gamma distribution has *mgf* $(1 - \theta z)^{-\alpha}$. Adding n such variables produces a random variable with *mgf* $(1 - \theta z)^{-n\alpha}$. This is a gamma distribution with parameters $n\alpha$ and θ. The inverse Gaussian distribution has *mgf* $\exp\{\theta[1 - (1 - 2\mu^2 z/\theta)^{1/2}]/\mu\}$. When raised to the nth power it becomes $\exp\{n\theta[1 - (1 - 2\mu^2 z/\theta)^{1/2}]/\mu\}$. This is an inverse Gaussian distribution with parameters $\mu^3/(n\theta^2)$ and μ^2/θ replacing μ and θ. Because both parameters change in the inverse Gaussian case, only the gamma has the desired property. However, when the inverse Gaussian distribution is reparameterized with $\beta = \mu^2/\theta$ replacing θ, the property will hold.

(b) All infinitely divisible distributions must be closed under convolutions. The answer is the same as that for Exercise 4.3. In (3.27), f_k^{*n} can be written explicitly. Note that this is the secondary distribution and so will not work for distributions such as the Poisson-ETNB in which the secondary distribution is truncated and so is not closed under convolution.

4.27 Since this is a compound distribution defined on the non-negative integers, we can use Theorem 3.6. With an appropriate adaptation of notation,

$$P_N[P_X(z; \beta)] = P_N\{P_X(z); \beta[1 - f_X(0)]\}.$$

So just replace β by $\beta^* = \beta[1 - f_X(0)]$.

4.28 (a)
$$f_S(x) = \sum_{n=1}^{\infty} \frac{\beta^n}{n(1 + \beta)^n \log(1 + \beta)} \frac{x^{n-1}e^{-x/\theta}}{\theta^n(n - 1)!}$$

$$= \frac{1}{\log(1 + \beta)} \sum_{n=1}^{\infty} \left[\frac{\beta}{\theta(1 + \beta)}\right]^n \frac{1}{n!} x^{n-1} e^{-x/\theta}$$

(b)
$$f_S(x) = \frac{e^{-x/\theta}}{x \log(1 + \beta)} \sum_{n=1}^{\infty} \left[\frac{x\beta}{\theta(1 + \beta)}\right]^n \frac{1}{n!}$$

$$= \frac{e^{-x/\theta}}{x \log(1+\beta)} \left\{ \exp\left[\frac{x\beta}{\theta(1+\beta)} \right] - 1 \right\}$$

$$= \frac{\exp\left[-\frac{x}{\theta(1+\beta)} \right] - \exp\left(-\frac{x}{\theta} \right)}{x \log(1+\beta)}$$

4.29 The answer is the sum of
$0.80R_{100}$ pays 80% of all in excess of 100,
$0.10R_{1100}$ pays an additional 10% in excess of 1,100, and
$0.10R_{2100}$ pays an additional 10% in excess of 2,100.

4.30 $\Pr(S = 0) = \sum_{n=0}^{\infty} p_n \Pr(B_n = 0)$ where $B_n \tilde{\,} bin(n, 0.25)$. Thus

$$\Pr(S = 0) = \sum_{n=0}^{\infty} \frac{e^{-2} 2^n}{n!} (0.75)^n = e^{-2} e^{1.5} = e^{-1/2}.$$

4.31
$$\begin{aligned}
E(S) &= \int_0^{\infty} x f(x) dx = \int_0^{\infty} x \int_2^4 f(x|\theta) u(\theta) d\theta \, dx \\
&= \int_2^4 \frac{1}{2} \int_0^{\infty} x f(x|\theta) dx \, d\theta = \frac{1}{2} \int_2^4 E(S|\theta) d\theta = \frac{1}{2} \int_2^4 \frac{1}{\theta} d\theta \\
&= \frac{1}{2} (\log 4 - \log 2)
\end{aligned}$$

4.32 $E(N) = 0.1$.

$$\begin{aligned}
E(X) &= \int_0^{30} 100t f(t) dt + 3,000 \Pr(T > 30) \\
&= \int_0^{10} 100t(0.04) dt + \int_{10}^{20} 100t(0.035) dt \\
&\quad + \int_{20}^{30} 100t(0.02) dt + 3,000(0.05) \\
&= 200 + 525 + 500 + 150 = 1,3\tilde{\,}0.
\end{aligned}$$

$E(S) = 0.1(1,350) = 135$.

4.33 Total claims are compound Poisson with $\lambda = 5$ and severity distribution

$$\begin{aligned}
f_X(x) &= 0.4 f_1(x) + 0.6 f_2(x) \\
&= \begin{cases} 0.4(0.001) + 0.6(0.005) = 0.0034, & 0 < x \le 200 \\ 0.4(0.001) + 0.6(0) = 0.0004, & 200 < x \le 1,000. \end{cases}
\end{aligned}$$

Then,

$$E[(X - 100)_+] = \int_{100}^{\infty} (x - 100) f_X(x) dx$$

$$= \int_{100}^{200} (x - 100)(0.0034) dx$$

$$+ \int_{200}^{\infty} (x - 100)(0.0004) dx$$

$$= 177.$$

4.34 The calculations appear in Table 3.5.

Table 3.5 Calculations for Exercise 4.34

s	$\Pr(S = s)$	d	$d \Pr(S = s)$
0	0.031676	4	0.12671
1	0.126705	3	0.38012
2	0.232293	2	0.46459
3	0.258104	1	0.25810
4+	0.351222	0	0
	Total		1.25952

4.35 (a)

$$M_X(z) = \sum_{k=1}^{r} q_k \int_0^{\infty} e^{zx} \theta^{-k} x^{k-1} e^{-x/\theta} \frac{1}{(n-1)!} dx$$

$$= \sum_{k=1}^{r} q_k (1 - \theta z)^{-k} = Q[(1 - \theta z)^{-1}]$$

(b)

$$M_S(z) = P_N[M_X(z)] = P_N\{Q[(1 - \theta z)^{-1}]\}$$
$$= C[(1 - \theta z)^{-1}]$$

where $C(z) = P_N[Q(z)]$.

(c) Use Theorem 3.4 with f_j replaced by c_j and g_j replaced by $f_X(j)$.

(d) S has a compound distribution with "frequency" pf c_k and "severity" mgf $(1 - \theta z)^{-1}$ which corresponds to an exponential distribution with parameter θ. The cdf is as in (4.11) with p_n replaced by c_n. Then $\{c_k\}$ is a compound distribution with N primary and $\{q_k\}$ secondary.

3.5 SECTION 4.6

4.36

$$
\begin{aligned}
m_0^k &= \int_{kh}^{(k+1)h} \frac{x - kh - h}{-h} f(x)\,dx \\
&= -\int_0^{(k+1)h} \frac{x}{h}\,dx + \int_0^{kh} \frac{x}{h}\,dx + (k+1)\{F[(k+1)h] - F(kh)\} \\
&= -\frac{1}{h}E[X \wedge (k+1)h] + (k+1)\{1 - F[(k+1)h]\} \\
&\quad +\frac{1}{h}E(X \wedge kh) - k[1 - F(kh)] \\
&\quad +(k+1)\{F[(k+1)h] - F(kh)\} \\
&= \frac{1}{h}E(X \wedge kh) - \frac{1}{h}E[X \wedge (k+1)h] + 1 - F(kh)
\end{aligned}
$$

$$
\begin{aligned}
m_1^k &= \int_{kh}^{(k+1)h} \frac{x - kh}{h} f(x)\,dx \\
&= \frac{1}{h}E[X \wedge (k+1)h] - \frac{1}{h}E(X \wedge kh) - 1 + F[(k+1)h]
\end{aligned}
$$

For $k = 1, 2, \ldots$,

$$
\begin{aligned}
f_k &= m_1^{k-1} + m_0^k \\
&= \frac{1}{h}E(X \wedge kh) - \frac{1}{h}E[X \wedge (k-1)h] - 1 + F(kh) \\
&\quad +\frac{1}{h}E(X \wedge kh) - \frac{1}{h}E[X \wedge (k+1)h] + 1 - F(kh) \\
&= \frac{1}{h}\{2E(X \wedge kh) - E[X \wedge (k-1)h] - E[X \wedge (k+1)h]\}.
\end{aligned}
$$

Also, $f_0 = m_0^0 = 1 - E(X \wedge h)/h$. All of the m_j^k have non-negative integrands and therefore all of the f_k are non-negative. To be a valid probability function they must add to one:

$$
\begin{aligned}
f_0 + \sum_{k=1}^{\infty} f_k &= 1 - \frac{1}{h}E(X \wedge h) + \frac{1}{h}\sum_{k=1}^{\infty}\{E(X \wedge kh) - E[X \wedge (k-1)h]\} \\
&\quad +\frac{1}{h}\sum_{k=1}^{\infty}\{E(X \wedge kh) - E[X \wedge (k+1)h]\} \\
&= 1 - \frac{1}{h}E(X \wedge h) + \frac{1}{h}E(X \wedge h) = 1
\end{aligned}
$$

because both sums are telescoping.

The mean of the discretized distribution is

$$\sum_{k=1}^{\infty} hkf_k = \sum_{k=1}^{\infty} k\{E(X \wedge kh) - E[X \wedge (k-1)h]\}$$

$$+ \sum_{k=1}^{\infty} k\{E(X \wedge kh) - E[X \wedge (k+1)h]\}$$

$$= E(X \wedge h) + \sum_{k=1}^{\infty} (k+1)\{E[X \wedge (k+1)h] - E(X \wedge (kh))\}$$

$$+ \sum_{k=1}^{\infty} k\{E(X \wedge kh) - E[X \wedge (k+1)h]\}$$

$$= E(X \wedge h) + \sum_{k=1}^{\infty} \{E[X \wedge (k+1)h] - E(X \wedge (kh))\}$$

$$= E(X)$$

becasue $E(X \wedge \infty) = E(X)$.

4.37 Assume $x = 1$. Then $g_0 = \exp[-200(1 - 0.76)] = \exp(-48)$. The recursive formula gives

$$g_k = \frac{200}{k}(0.14g_{k-1} + 0.10g_{k-2} + 0.06g_{k-3} + 0.12g_{k-4})$$

with $g_k = 0$ for $k < 0$. Now use a spreadsheet to recursively compute probabilities until the probabilities sum to 0.05. This happens at $k = 62$. Then $62x = 4,000,000$ for $x = 64,516$. The expected compensation is

$$200(0.14 + 0.10 + 0.06 + 0.12)(64,516) = 5,419,344.$$

4.38 (a) $P_N(z) = wP_1(z) + (1-w)P_2(z)$
(b)
$$\begin{aligned}
P_S(z) &= P_N(P_X(z)) \\
&= wP_1(P_X(z)) + (1-w)P_2(P_X(z)) \\
&= wP_{S_1}(z) + (1-w)P_{S_2}(z)
\end{aligned}$$

$$f_S(x) = wf_{S_1} + (1-w)f_{S_2}(x), \quad x = 0, 1, 2, \dots$$

Hence, first use (4.18) to compute $f_{S_1}(x)$. Then take a weighted average of the results.
(c) Yes. Any distributions $P_1(z)$ and $P_2(z)$ using (4.18) can be used.

4.39 From (4.18), the recursion for the compound Poisson distribution,

$$f_S(0) = e^{-5}$$

$$f_S(x) = \frac{5}{x} \sum_{y=1}^{5} y f_X(y) f_S(x - y).$$

Then

$$f_S(1) = 5 f_X(1) e^{-5}$$

and so $f_X(1) = \frac{1}{5}$ since $f_S(1) = e^{-5}$.

$$
\begin{aligned}
f_S(2) &= \frac{5}{2}[f_X(1)f_S(1) + 2f_X(2)f_S(0)] \\
&= \frac{5}{2}[\frac{1}{5}e^{-5} + 2f_X(2)e^{-5}]
\end{aligned}
$$

since $f_S(2) = \frac{5}{2}e^{-5}$, we obtain $f_X(2) = \frac{2}{5}$.

4.40 $f_S(7) = \frac{6}{7}[1 f_X(1)f_S(6) + 2f_X(2)f_S(5) + 4f_X(4)f_S(3)]$. Therefore, $0.041 = \frac{6}{7}[\frac{1}{3}f_S(6) + \frac{2}{3}0.0271 + \frac{4}{3}0.0132]$ for $f_S(6) = 0.0365$.

4.41 From (4.17) with $f_X(0) = 0$ and x replaced by z:

$$
\begin{aligned}
f_S(z) &= \sum_{y=1}^{M} \left(a + b\frac{y}{z}\right) f_X(y) f_S(z - y) \\
&= \sum_{y=1}^{M-1} \left(a + b\frac{y}{z}\right) f_X(y) f_S(z - y) \\
&\quad + \left(a + b\frac{M}{z}\right) f_X(M) f_S(z - M).
\end{aligned}
$$

Let $z = x + M$

$$
\begin{aligned}
f_S(x + M) &= \sum_{y=1}^{M-1} \left(a + b\frac{y}{x + M}\right) f_X(y) f_S(x + M - y) \\
&\quad + \left(a + b\frac{M}{x + M}\right) f_X(M) f_S(x)
\end{aligned}
$$

Rearrangement gives the result.

4.42 We are looking for $6 - E(S) - E(D)$.
$E(X) = \frac{1}{4}1 + \frac{3}{4}2 = \frac{7}{4}$. $E(S) = 2\frac{7}{4} = \frac{7}{2}$.

$$D = \begin{cases} 4.5 - S, & S < 4.5 \\ 0, & S \geq 4.5. \end{cases}$$

$$f_S(0) = e^{-2},$$

$$f_S(1) = \frac{2}{1}1\frac{1}{4}e^{-2} = \frac{1}{2}e^{-2},$$

$$f_S(2) = \frac{2}{2}\left(1\frac{1}{4}\frac{1}{2}e^{-2} + 2\frac{3}{4}e^{-2}\right) = \frac{13}{8}e^{-2},$$

$$f_S(3) = \frac{2}{3}\left(1\frac{1}{4}\frac{13}{8}e^{-2} + 2\frac{3}{4}\frac{1}{2}e^{-2}\right) = \frac{37}{48}e^{-2},$$

$$f_S(4) = \frac{2}{4}\left(1\frac{1}{4}\frac{37}{48}e^{-2} + 2\frac{3}{4}\frac{13}{8}e^{-2}\right) = \frac{505}{384}e^{-2}.$$

Then $E(D) = (4.5 + 3.5\frac{1}{2} + 2.5\frac{13}{8} + 1.5\frac{37}{48} + 0.5\frac{505}{384})e^{-2} = 1.6411$. The answer is $6 - 3.5 - 1.6411 = 0.8589$.

4.43 For adults, the distribution is compound Poisson with $\lambda = 3$ and severity distribution with probabilities 0.4 and 0.6 on 1 and 2 (in units of 200). For children it is $\lambda = 2$ and severity probabilities of 0.9 and 0.1. The sum of the two distributions is also compound Poisson, with $\lambda = 5$. The probability at 1 is $[3(0.4) + 2(0.9)]/5 = 0.6$ and the remaining probability is at 2. The initial aggregate probabilities are:

$$f_S(0) = e^{-5},$$

$$f_S(1) = \frac{5}{1}1\frac{3}{5}e^{-5} = 3e^{-5},$$

$$f_S(2) = \frac{5}{2}\left(1\frac{3}{5}3e^{-5} + 2\frac{2}{5}e^{-5}\right) = \frac{13}{2}e^{-5},$$

$$f_S(3) = \frac{5}{3}\left(1\frac{3}{5}\frac{13}{2}e^{-5} + 2\frac{2}{5}3e^{-5}\right) = \frac{21}{2}e^{-5},$$

$$f_S(4) = \frac{5}{4}\left(1\frac{3}{5}\frac{21}{2}e^{-5} + 2\frac{2}{5}\frac{13}{2}e^{-5}\right) = \frac{115}{8}e^{-5}.$$

The probability of claims being 800 or less is the sum

$$(1 + 3 + \frac{13}{2} + \frac{21}{2} + \frac{115}{8})e^{-5} = 35.375e^{-5} = 0.2384.$$

4.44 The aggregate distribution is two times a Poisson variable. The probabilities are $\Pr(S = 0) = e^{-1}$, $\Pr(S = 2) = \Pr(N = 1) = e^{-1}$, $\Pr(S = 4) = \Pr(N = 2) = \frac{1}{2}e^{-1}$. $E(D) = (6-0)e^{-1} + (6-2)e^{-1} + (6-4)\frac{1}{2}e^{-1} = 11e^{-1} = 4.0467$.

4.45 $\lambda = 1 + 1 = 2$, $f_X(1) = [1(1) + 1(0.5)]/2 = 0.75$, $f_X(2) = 0.25$. The calculations appear in Table 3.6. The answer is $F_X^{*4}(6) = (81+108+54)/256 = 243/256 = 0.94922$.

Table 3.6   Calculations for Exercise 4.50

x	$f_X^{*0}(x)$	$f_X^{*1}(x)$	$f_X^{*2}(x)$	$f_X^{*3}(x)$	$f_X^{*4}(x)$
0	1	0	0	0	0
1	0	3/4	0	0	0
2	0	1/4	9/16	0	0
3	0	0	6/16	27/64	0
4	0	0	1/16	27/64	81/256
5	0	0	0	9/64	108/256
6	0	0	0	1/64	54/256

4.46 $56 = 29E(X)$, so $E(X) = 56/29$. $126 = 29E(X^2)$, so $E(X^2) = 126/29$. Let $f_i = \Pr(X = i)$. Then there are three equations:
$f_1 + f_2 + f_3 = 1$
$f_1 + 2f_2 + 3f_3 = 56/29$
$f_1 + 4f_2 + 9f_3 = 126/29$.
The solution is $f_2 = 11/29$. (Also, $f_1 = 10/29$ and $f_3 = 8/29$)

4.47 Let $f_j = \Pr(X = j)$. $0.16 = \lambda f_1$, $k = 2\lambda f_2$, $0.72 = 3\lambda f_3$. Then $f_1 = 0.16/\lambda$ and $f_3 = 0.24/\lambda$ and so $f_2 = 1 - 0.16/\lambda - 0.24/\lambda$. $1.68 = \lambda[1(0.16/\lambda) + 2(1 - 0.4/\lambda) + 3(0.24/\lambda)] = 0.08 + 2\lambda$ and so $\lambda = 0.8$.

4.48 $1 - F(6) = 0.04 - 0.02$, $F(6) = 0.98$. $1 - F(4) = 0.20 - 0.10$, $F(4) = 0.90$. $\Pr(S = 5 \text{ or } 6) = F(6) - F(4) = 0.08$.

4.49 For 1,500 lives $\lambda = 0.01(1,500) = 15$. In units of 20, the severity distribution is $\Pr(X = 0) = 0.5$ and $\Pr(X = x) = 0.1$ for $x = 1, 2, 3, 4, 5$. Then $E(X^2) = 0.5(0) + 0.1(1 + 4 + 9 + 16 + 25) = 5.5$ and $Var(S) = 15(5.5) = 82.5$. In payment units it is $20^2(82.5) = 33,000$.

4.50 $\Pr(N = 2) = \frac{1}{4}\Pr(N = 2|\text{class I}) + \frac{3}{4}\Pr(N = 2|\text{class II})$. $\Pr(N = 2|\text{class}$ I$) = \int_0^1 \frac{\beta^2}{(1+\beta)^3} 1 d\beta = 0.068147$. (Hint: Use the substitution $y = \beta/(1 + \beta)$). $\Pr(N = 2|\text{class II}) = (0.25)^2/(1.25)^3 = 0.032$. $\Pr(N = 2) = 0.25(0.068147) + 0.75(0.32) = 0.04104$.

4.51 Using program CR, the result is 0.016.

4.52 Using program CR, the result is 0.055. Using a normal approximation with a mean of 250 and a variance 8,000 the probability is 0.047.

3.6   SECTION 4.7

4.53 The answers are the same as for Exercises 4.51 and 4.52.

3.7   SECTION 4.8

4.54 The answers are about the same as for Exercises 4.51 and 4.52.

3.8   SECTION 4.9

4.55 With a deductible of 25 the probability of making a payment is $v = 1 - F(25) = 0.98101$. The frequency distribution for the number of payments remains negative binomial with $r = 2$ and $\beta = 0.98101(2) = 1.96202$. The discretized severity distribution for payments begins with $f_0 = [F(27.5) - F(25)]/[1 - F(25)] = 0.00688$ and $f_1 = [F(32.5) - F(27.5)]/[1 - F(25)] = 0.01770$. The first 41 values of the discretized distribution and the discretized aggregate distribution are given in Table 3.7. The estimate value of $F(200)$ is obtained by summing all but the last given aggregate probabilities and then adding half of the final one. The result is 0.44802. For the limited expected value use (4.32) to obtain 146.3925. We also have $g_0 = (1+\beta)^{-r} = 0.11398$.

For the disappearing deductible, the same negative binomial distribution is used. To discretize the severity distribution we have

$$
\begin{aligned}
f_0 &= \Pr(Y \leq 2.5) = \Pr\left[\frac{200(X - 25)}{175} \leq 2.5\right] \\
&= \Pr(25 < X \leq 27.1875) \\
f_1 &= \Pr(2.5 < X \leq 7.5) = \Pr\left[2.5 < \frac{200(X - 25)}{175} \leq 7.5\right] \\
&= \Pr(27.1875 < X \leq 31.5625) \\
&\qquad\vdots \\
f_{39} &= \Pr(192.5 < Y \leq 197.5) = \Pr\left[192.5 < \frac{200(X - 25)}{175} \leq 197.5\right] \\
&= \Pr(193.4375 < X \leq 197.8125) \\
f_{40} &= \Pr(197.5 < Y \leq 202.5) = \Pr(197.8125 < X \leq 202.5). \\
f_{41} &= \Pr(202.5 < X \leq 207.5) \\
&\qquad\vdots
\end{aligned}
$$

The values appear in Table 3.8 and $F(200) = 0.40485$ and $E(X \wedge 200) = 150.9259$.

Table 3.7   Discretized severities and aggregate probabilites for an ordinary deductible

x	f_x	p_x	x	f_x	p_x
0	0.006884	0.115025			
5	0.017690	0.002708	105	0.026359	0.009622
10	0.022983	0.003566	110	0.024169	0.009589
15	0.028080	0.004424	115	0.022070	0.009549
20	0.032730	0.005246	120	0.020076	0.009505
25	0.036757	0.006008	125	0.018198	0.009456
30	0.040056	0.006696	130	0.016440	0.009403
35	0.042579	0.007301	135	0.014805	0.009347
40	0.044326	0.007822	140	0.013294	0.009289
45	0.045329	0.008260	145	0.011904	0.009228
50	0.045650	0.008622	150	0.010632	0.009165
55	0.045363	0.008914	155	0.009473	0.009101
60	0.044551	0.009145	160	0.008420	0.009034
65	0.043302	0.009322	165	0.007469	0.008965
70	0.041698	0.009453	170	0.006610	0.008895
75	0.039821	0.009547	175	0.005839	0.008823
80	0.037744	0.009609	180	0.005149	0.008750
85	0.035533	0.009645	185	0.004531	0.008675
90	0.033244	0.009661	190	0.003981	0.008598
95	0.030926	0.009660	195	0.003493	0.008520
100	0.028620	0.009646	200	0.003059	0.008441

4.56 (a) The recursive formula was used with a discretization interval of 25 and a mean-preserving discretization. The answers are 6,192.69, 4,632.13, and 12,800.04.

(b) The individual deductible of 100 requires a change in the frequency distribution. The probability of exceeding 100 under the gamma distribution is 0.7772974 and so the new Poisson parameter is 5 times this probability or 3.886487. The results are 5,773.24, 4,578.78, and 12,073.35.

(c) The frequency distribution is altered as in part (b). The results are 148.27, 909.44, and 0.

4.57 (a) Two passes of the recursive formula were used, first with the Poisson(4) distribution and the two point severity distribution. The second pass uses the output from the first pass as the severity distribution along with a Poisson(10) frequency. With an aggregate limit of 400 the results are 247.25, 83.20, and 368.34.

(b) The mean is found from $0.8[E(X) - E(X \wedge 300)] = 13.78$. The other results are 31.90 and 54.67.

Table 3.8   Discretized severities and aggregate probabilites for a disappearing deductible

x	f_x	p_x	x	f_x	p_x
0	0.005946	0.114882			
5	0.014897	0.002276	105	0.028330	0.008455
10	0.018957	0.002930	110	0.026554	0.008450
15	0.022934	0.003591	115	0.024793	0.008438
20	0.026667	0.004235	120	0.023065	0.008419
25	0.030034	0.004847	125	0.021384	0.008394
30	0.032951	0.005414	130	0.019762	0.008365
35	0.035368	0.005929	135	0.018210	0.008332
40	0.037263	0.006389	140	0.016732	0.008295
45	0.038636	0.006791	145	0.015333	0.008257
50	0.039508	0.007138	150	0.014017	0.008216
55	0.039910	0.007433	155	0.012783	0.008173
60	0.039885	0.007679	160	0.011632	0.008128
65	0.039481	0.007881	165	0.010562	0.008081
70	0.038746	0.008044	170	0.009572	0.008034
75	0.037733	0.008173	175	0.008658	0.007984
80	0.036489	0.008271	180	0.007817	0.007934
85	0.035062	0.008345	185	0.007045	0.007882
90	0.033496	0.008396	190	0.006339	0.007829
95	0.031829	0.008430	195	0.005695	0.007775
100	0.030097	0.008448	200	0.005456	0.007773

3.9   SECTION 4.12

4.60
$$\begin{aligned} B &= E(S) + 2SD(S) \\ &= 40(2) + 60(4) + 2\sqrt{40(4) + 60(10)} \\ &= 375.136. \end{aligned}$$

For A,
$$\begin{aligned} E(S) &= E[E(S|N)] = E[2N + 4(100 - N)] \\ &= 400 - 2E(N) = 400 - 2(40) = 320. \end{aligned}$$

$$\begin{aligned} Var(S) &= Var[E(S|N)] + E[Var(S|N)] \\ &= Var(400 - 2N) + E[4N + 10(100 - N)] \\ &= 4Var(N) - 6E(N) + 1000 \\ &= 4(100)(0.4)(0.6) - 6(100)(0.4) + 1000 = 856. \end{aligned}$$

Therefore, $A = 320 + 2\sqrt{856} = 378.515$ and $A/B = 1.009$.

4.61 Premium per person is $1.1(1,000)[0.2(0.02) + 0.8(0.01)] = 13.20$. With 30% smokers,

$$
\begin{aligned}
E(S) &= 1,000[0.3(0.02) + 0.7(0.01)] = 13, \\
Var(S) &= 1,000^2[0.3(0.02)(0.98) + 0.7(0.01)0(.99)] = 12,810.
\end{aligned}
$$

With n policies, the probability of claims exceeding premium is

$$
\begin{aligned}
\Pr(S > 13.2n) &= \Pr\left(Z > \frac{13.2n - 13n}{\sqrt{12,810n}}\right) \\
&= \Pr(Z > .0017671\sqrt{n}) = 0.20.
\end{aligned}
$$

Therefore, $0.0017671\sqrt{n} = 0.84162$ for $n = 226,836$.

4.62 Let the policy being changed be the nth policy. Then, originally, $F_S^{*n}(x) = 0.8F_S^{*(n-1)}(x) + 0.2F_S^{*(n-1)}(x-1)$. Starting with $x = 0$:

$$
\begin{aligned}
0.40 &= 0.80F_S^{*(n-1)}(0) + 0.2(0), & F_S^{*(n-1)}(0) &= 0.50 \\
0.58 &= 0.80F_S^{*(n-1)}(1) + 0.2(0.50), & F_S^{*(n-1)}(1) &= 0.60 \\
0.64 &= 0.80F_S^{*(n-1)}(2) + 0.2(0.60), & F_S^{*(n-1)}(2) &= 0.65 \\
0.69 &= 0.80F_S^{*(n-1)}(3) + 0.2(0.65), & F_S^{*(n-1)}(3) &= 0.70 \\
0.70 &= 0.80F_S^{*(n-1)}(4) + 0.2(0.70), & F_S^{*(n-1)}(4) &= 0.70 \\
0.78 &= 0.80F_S^{*(n-1)}(5) + 0.2(0.70), & F_S^{*(n-1)}(5) &= 0.80
\end{aligned}
$$

With the amount changed, $F_S^{*n}(5) = 0.8F_S^{*(n-1)}(5) + 0.2F_S^{*(n-1)}(3) = 0.8(0.8) + 0.2(0.7) = 0.78$.

4.63 $E(S) = 400(0.03)(5) + 300(0.07)(3) + 200(0.10)(2) = 163$.
For a single insured with claim probability q and exponential mean θ,

$$
\begin{aligned}
Var(S) &= E(N)Var(X) + Var(N)E(X)^2 \\
&= q\theta^2 + q(1-q)\theta^2 \\
&= q(2-q)\theta^2.
\end{aligned}
$$

$$
\begin{aligned}
Var(S) &= 400(0.03)(1.97)(25) + 300(0.07)(1.93)(9) \\
&\quad + 200(0.10)(1.90)(4) \\
&= 1,107.77
\end{aligned}
$$

$P = E(S) + 1.645SD(S) = 163 + 1.645\sqrt{1,107.77} = 217.75$.

4.66 $\lambda = 500(0.01) + 500(0.02) = 15$. $f_X(x) = 500(0.01)/15 = 1/3$, $f_X(2x) = 2/3$. $E(X^2) = (1/3)x^2 + (2/3)(2x)^2 = 3x^2$. $Var(S) = 15(3x^2) = 45x^2 = 4,500$. $x = 10$.

4.67 All work is done in units of 100,000. The first group of 500 policies is not relevant. The others have amounts 1, 2, 1, and 1.

$$
\begin{aligned}
E(S) &= 500(0.02)(1) + 500(0.02)(2) + 300(0.1)(1) + 500(0.1)(1) \\
&= 110 \\
Var(S) &= 500(0.02)(0.98)(1) + 500(0.02)(0.98)(4) \\
&\quad + 300(0.1)(0.9)(1) + 500(0.1)(0.9)(1) \\
&= 121
\end{aligned}
$$

(a) $P = 110 + 1.645\sqrt{121} = 128.095$.

(b) $\mu + \sigma^2/2 = \log 110 = 4.70048$. $2\mu + 2\sigma^2 = \log(121 + 110^2) = 9.41091$. $\mu = 4.695505$ and $\sigma = .0997497$. $\log P = 4.695505 + 1.645(0.0997497) = 4.859593$, $P = 128.97$.

(c) $\alpha\theta = 110$, $\alpha\theta^2 = 121$, $\theta = 121/110 = 1.1$, $\alpha = 100$. This is a chi-square distribution with 200 degrees of freedom. The 95th percentile is (using the Excel GAMMAINV function) 128.70.

(d) $\lambda = 500(0.02) + 500(0.02) + 300(0.10) + 500(0.10) = 100$. $f_X(1) = (10 + 30 + 50)/100 = 0.9$ and $f_X(2) = .1$. Using the recursive formula for compound Poisson distribution (for example, with program CR), we find $F_S(128) = 0.94454$ and $F_S(129) = 0.95320$ and so to be safe at the 5% level a premium of 129 is required.

(e) One way to use the software is to note that $S = X_1 + 2X_2 + X_3$ where each X is binomial with $m = 500$, 500, and 800 and $q = 0.02$, 0.02, and 0.10. The three densities can be created in program CR and then the convolutions done with program CRC2. The results are $F_S(128) = 0.94647$ and $F_S(129) = 0.95635$ and so to be safe at the 5% level a premium of 129 is required.

4

Chapter 5 Solutions

4.1 SECTION 5.2

5.1
$$f(x|y) = \frac{f(x,y)}{f(y)} = \frac{\Pr(X=x,\ Z=y-x)}{\Pr(Y=y)}$$
$$= \frac{\binom{n_1}{x}p^x(1-p)^{n_1-x}\binom{n_2}{y-x}p^{y-x}(1-p)^{n_2-y+x}}{\binom{n_1+n_2}{y}p^y(1-p)^{n_1+n_2-y}}$$
$$= \frac{\binom{n_1}{x}\binom{n_2}{y-x}}{\binom{n_1+n_2}{y}}$$

This is the hypergeometric distribution.

5.2(a) $f_X(0) = 0.3,\ f_X(1) = 0.4,\ f_X(2) = 0.3$.
 $f_Y(0) = 0.25,\ f_Y(1) = 0.3,\ f_Y(2) = 0.45$.
(b) The following array presents the values for $x = 0, 1, 2$

$$
\begin{aligned}
f(x|Y=0) &= 0.2/0.25 = 0.8,\ 0/0.25 = 0,\ 0.05/0.25 = 0.2,\\
f(x|Y=1) &= 0/0.3 = 0,\ 0.15/0.3 = 0.5,\ 0.15/0.3 = 0.5,\\
f(x|Y=2) &= 0.1/0.45 = 0.22,\ 0.25/0.45 = 0.56,\ 0.1/0.45 = 0.22
\end{aligned}
$$

(c)
$$
\begin{aligned}
E(X|Y=0) &= 0(0.8) + 1(0) + 2(0.2) = 0.4\\
E(X|Y=1) &= 0(0) + 1(0.5) + 2(0.5) = 1.5\\
E(X|Y=2) &= 0(0.22) + 1(0.56) + 2(0.22) = 1\\
E(X^2|Y=0) &= 0(0.8) + 1(0) + 4(0.2) = 0.8
\end{aligned}
$$

$$
\begin{aligned}
E(X^2|Y=1) &= 0(0) + 1(0.5) + 4(0.5) = 2.5 \\
E(X^2|Y=2) &= 0(0.22) + 1(0.56) + 4(0.22) = 1.44 \\
Var(X|Y=0) &= 0.8 - 0.4^2 = 0.64 \\
Var(X|Y=1) &= 2.5 - 1.5^2 = 0.25 \\
Var(X|Y=2) &= 1.44 - 1^2 = 0.44
\end{aligned}
$$

(d)
$$
\begin{aligned}
E(X) &= 0.4(0.25) + 1.5(0.3) + 1(0.45) = 1 \\
E[Var(X|Y)] &= 0.64(0.25) + 0.25(0.3) + 0.44(0.45) = 0.433 \\
Var[E(X|Y)] &= 0.16(0.25) + 2.25(0.3) + 1(0.45) - 1^2 = 0.165 \\
Var(X) &= 0.433 + 0.165 = 0.598
\end{aligned}
$$

5.3 (a)

$$
f(x,y) \propto \exp\left\{ -\frac{1}{2(1-\rho^2)} \left[\left(\frac{x-\mu_1}{\sigma_1}\right)^2 - 2\rho\left(\frac{x-\mu_1}{\sigma_1}\right)\left(\frac{y-\mu_2}{\sigma_2}\right) \right] \right\}
$$

$$
\propto \exp\left\{ -\frac{1}{2(1-\rho^2)} \left[\frac{x^2}{\sigma_1^2} - 2x\left(\frac{\mu_1}{\sigma_1^2} + \rho\frac{y-u_2}{\sigma_1\sigma_2}\right) \right] \right\}.
$$

Now a normal density $N(\mu, \sigma^2)$ has *pdf* $f(x) \propto \exp\left[-\frac{1}{2\sigma^2}(x^2 - 2\mu x)\right]$. Then $f_{X|Y}(x|y) \propto f(x,y)$ is $N\left[\mu_1 + \rho\frac{\sigma_1}{\sigma_2}(y-\mu_2), (1-\rho^2)\sigma_1^2\right]$.

(b)

$$
\begin{aligned}
f_X(x) &= \int f(x,y)dy \\
&\propto \int \exp\left\{ -\frac{1}{2(1-\rho^2)} \left[\left(\frac{x-\mu_1}{\sigma_1}\right)^2 - 2\rho\left(\frac{x-\mu_1}{\sigma_1}\right)\left(\frac{y-\mu_2}{\sigma_2}\right) \right.\right. \\
&\qquad \left.\left. + \left(\frac{y-\mu_2}{\sigma_2}\right)^2 \right] \right\} dy \\
&= \exp\left[-\frac{1}{2}\left(\frac{x-\mu_1}{\sigma_1}\right)^2 \right] \\
&\qquad \times \int \exp\left[-\frac{1}{2}\left(\frac{y-\mu_2-\rho\frac{\sigma_2}{\sigma_1}(x-\mu_1)}{\sigma_2\sqrt{1-\rho^2}}\right)^2 \right] dy \\
&= \exp\left[-\frac{1}{2}\left(\frac{x-\mu_1}{\sigma_1}\right)^2 \right] \sqrt{2\pi\sigma_2\sqrt{1-\rho^2}} \\
&\propto \exp\left[-\frac{1}{2}\left(\frac{x-\mu_1}{\sigma_1}\right)^2 \right]
\end{aligned}
$$

Since the normal density is $\frac{1}{\sqrt{2\pi}\sigma}\exp\left[-\frac{1}{2}\left(\frac{x-\mu}{\sigma}\right)^2\right]$, in general, we have

$$f_X(x) = \frac{1}{\sqrt{2\pi}\sigma_1}\exp\left[-\frac{1}{2}\left(\frac{x-\mu_1}{\sigma_1}\right)^2\right] \sim N\left(\mu_1,\sigma_1^2\right)$$

(c) Suppose $f_X(x)f_Y(y) = f_{X,Y}(x,y)$. Then $f_X(x)f_Y(y) = f_{X,Y}(x,y) = f_{X|Y}(x|y)f_Y(y)$ Therefore, $f_X(x) = f_{X|Y}(x|y)$. From the results of (a) and (b), $\rho = 0$.

Then

$$f_{X,Y}(x,y) \propto \exp\left\{-\frac{1}{2}\left[\left(\frac{x-\mu_1}{\sigma_1}\right)^2 + \left(\frac{y-\mu_2}{\sigma_2}\right)^2\right]\right\} \propto f_X(x)f_Y(x).$$

Therefore, $f_{X,Y}(x,y) = f_X(x)f_Y(y)$.

5.4
$$\begin{aligned}
E\left[\sum b_j(Y_j - \bar{Y})^2\right] &= E\left[\sum b_j(Y_j - \gamma + \gamma - \bar{Y})^2\right] \\
&= E\left[\sum b_j(Y_j - \gamma)^2 + \sum b_j(\gamma - \bar{Y})^2\right. \\
&\quad \left. +2\sum b_j(Y_j - \gamma)(\gamma - \bar{Y})\right] \\
&= \sum b_j(a_j + \sigma^2/b_j) + bVar(\bar{Y}) \\
&\quad +2E\left[\gamma\sum b_jY_j - b\gamma^2 + b\gamma\bar{Y} - \bar{Y}\sum b_jY_j\right]
\end{aligned}$$

$$\begin{aligned}
E\left[\gamma\sum b_jY_j - b\gamma^2 + b\gamma\bar{Y} - \bar{Y}\sum b_jY_j\right] &= E[\gamma b\bar{Y} - b\gamma^2 + b\gamma\bar{Y} - b\bar{Y}^2] \\
&= -E[b(\bar{Y} - \gamma)^2] \\
&= -bVar(\bar{Y})
\end{aligned}$$

$$\begin{aligned}
Var(\bar{Y}) &= \frac{1}{b^2}\sum b_j^2 Var(Y_j) = \frac{1}{b^2}\sum b_j^2(a_j + \sigma^2/b_j) \\
&= \frac{1}{b^2}\sum b_j^2 a_j + \frac{1}{b}\sigma^2
\end{aligned}$$

$$\begin{aligned}
E\left[\sum b_j(Y_j - \bar{Y})^2\right] &= \sum b_j(a_j + \sigma^2/b_j) - bVar(\bar{Y}) \\
&= \sum b_j(a_j + \sigma^2/b_j) - \frac{1}{b}\sum b_j^2 a_j - \sigma^2 \\
&= \sum a_j(b_j - b_j^2/b) + (n-1)\sigma^2
\end{aligned}$$

5.5 (a)
$$\sum\frac{(x_j - \mu)^2}{x_j} = \sum\left(x_j - 2\mu + \frac{\mu^2}{x_j}\right)$$

$$= \sum \left(\frac{\mu^2}{x_j} - \frac{\mu^2}{\bar{x}} \right) + \sum \left(\frac{\mu^2}{\bar{x}} - 2\mu + x_j \right)$$

$$= \mu^2 \sum \left(\frac{1}{x_j} - \frac{1}{\bar{x}} \right) + \frac{n\mu^2}{\bar{x}} - 2n\mu + n\bar{x}$$

$$= \mu^2 \sum \left(\frac{1}{x_j} - \frac{1}{\bar{x}} \right) + \frac{n}{\bar{x}} (\bar{x} - \mu)^2$$

(b)

$$L \propto \theta^{n/2} \exp \left[-\frac{\theta}{2\mu^2} \sum \frac{(x_j - \mu)^2}{x_j} \right]$$

$$l = \ln L = \frac{n}{2} \ln \theta - \frac{\theta}{2\mu^2} \sum \frac{(x_j - \mu)^2}{x_j}$$

$$= \frac{n}{2} \ln \theta - \frac{\theta}{2\mu^2} \left[\mu^2 \sum \left(\frac{1}{x_j} - \frac{1}{\bar{x}} \right) + \frac{n}{\bar{x}} (\bar{x} - \mu)^2 \right]$$

$$= \frac{n}{2} \ln \theta - \frac{\theta}{2} \sum \left(\frac{1}{x_j} - \frac{1}{\bar{x}} \right) - \frac{n\theta}{2\mu^2 \bar{x}} (\bar{x} - \mu)^2$$

$$\frac{\partial l}{\partial \mu} = -\frac{n\theta}{2\bar{x}} \frac{-\mu^2 2(\bar{x} - \mu) - (\bar{x} - \mu)^2 2\mu}{\mu^4} = 0$$

$$\hat{\mu} = \bar{x}$$

$$\frac{\partial l}{\partial \theta} = \frac{n}{2\theta} - \frac{1}{2} \sum \left(\frac{1}{x_j} - \frac{1}{\bar{x}} \right) + \frac{n}{2\mu^2 \bar{x}} (\bar{x} - \mu)^2 = 0$$

$$\hat{\theta} = \frac{n}{\sum \left(\frac{1}{x_j} - \frac{1}{\bar{x}} \right)}$$

5.6

$$L(\mu, \theta) = \prod_{j=1}^{n} f \left[x_j; \mu, (\theta m_j)^{-1} \right]$$

$$= \prod_{j=1}^{n} \left(\frac{2\pi}{\theta m_j} \right)^{-\frac{1}{2}} \exp \left[-\frac{(x_j - \mu)^2 m_j \theta}{2} \right]$$

$$\propto \theta^{n/2} \exp \left[-\frac{\theta}{2} \sum m_j (x_j - \mu)^2 \right]$$

$$\ell(\mu, \theta) = \frac{n}{2} \log \theta - \frac{\theta}{2} \sum m_j (x_j - \mu)^2 + \text{constant}$$

$$\frac{\partial \ell}{\partial \mu} = \theta \sum m_j (x_j - \mu) = 0 \Rightarrow \hat{\mu} = \frac{\sum m_j x_j}{\sum m_j} = \frac{\sum m_j x_j}{m}$$

$$\frac{\partial^2 \ell}{\partial \mu} = -\theta \sum m_j < 0, \text{ hence, maximum.}$$

$$\frac{\partial \ell}{\partial \theta} = \frac{n}{2} \frac{1}{\theta} - \frac{1}{2} \sum m_j (x_j - \mu)^2 = 0 \Rightarrow \hat{\theta}^{-1} = \frac{1}{n} \sum m_j (x_j - \hat{\mu})^2$$

$$\hat{\theta} = n\left[\sum m_j(x_j - \bar{x})^2\right]^{-1}$$

$$\frac{\partial^2 \ell}{\partial \theta^2} = -\frac{n}{2}\frac{1}{\theta^2} < 0, \text{ hence, maximum}$$

5.7 (a)
$$f(x) = \int_0^1 \binom{K}{x}\theta^x(1-\theta)^{K-x}\frac{\Gamma(a+b)}{\Gamma(a)\Gamma(b)}\theta^{a-1}(1-\theta)^{b-1}d\theta$$

$$= \binom{K}{x}\frac{\Gamma(a+b)}{\Gamma(a)\Gamma(b)}\frac{\Gamma(x+a)\Gamma(K-x+b)}{\Gamma(a+b+K)}$$

$$= \frac{K!}{x!(K-x)!}\frac{\Gamma(x+a)}{\Gamma(a)}\frac{\Gamma(K-x+b)}{\Gamma(b)}\frac{\Gamma(a+b)}{\Gamma(a+b+K)}$$

$$= \frac{a(a+1)\cdots(x+a-1)}{x!}\frac{b(b+1)\cdots(K-x+b-1)}{(K-x)!}$$

$$\times\frac{K!}{(a+b)\cdots(K+a+b-1)}$$

$$= \frac{(-1)^x\binom{-a}{x}(-1)^{K-x}\binom{-b}{K-x}}{(-1)^K\binom{-a-b}{K}} = \frac{\binom{-a}{x}\binom{-b}{K-x}}{\binom{-a-b}{K}}$$

$E(X|\theta) = K\theta$. $E(X) = E[E(X|\theta)] = E(K\theta) = K\frac{a}{a+b}$.

(b) $\pi(\theta|\mathbf{x}) \propto \theta^{\Sigma x_j}(1-\theta)^{\Sigma K_j - x_j}\theta^{a-1}(1-\theta)^{b-1}$ which is a beta density. Therefore the actual posterior distribution is

$$\pi(\theta|\mathbf{x}) = \frac{\Gamma(a+b+\Sigma K_j)}{\Gamma(a+\Sigma x_j)\Gamma(b+\Sigma K_j + \Sigma x_j)}\theta^{a+\Sigma x_j - 1}(1-\theta)^{b+\Sigma K_j - \Sigma x_j - 1}$$

with mean
$$E(\theta|\mathbf{x}) = \frac{a+\Sigma x_j}{a+b+\Sigma K_j}.$$

5.8 (a)
$$f(x) = \int_0^\infty \theta e^{-\theta x}\frac{\theta^{\alpha-1}e^{-\theta/\beta}}{\Gamma(\alpha)\beta^\alpha}d\theta$$

$$= \frac{1}{\Gamma(\alpha)\beta^\alpha}\int_0^\infty \theta^\alpha e^{-\theta(x+\beta^{-1})}d\theta$$

$$= \frac{1}{\Gamma(\alpha)\beta^\alpha}\Gamma(\alpha+1)(x+\beta^{-1})^{-\alpha-1}$$

$$= \alpha\beta^{-\alpha}(\beta^{-1}+x)^{-\alpha-1}$$

$E(X|\theta) = \theta^{-1}$.

$$E(X) = E[E(X|\theta)] = E(\theta^{-1})$$

$$= \int_0^\infty \theta^{-1}\frac{\theta^{\alpha-1}e^{-\theta/\beta}}{\Gamma(\alpha)\beta^\alpha}d\theta = \frac{\Gamma(\alpha-1)\beta^{\alpha-1}}{\Gamma(\alpha)\beta^\alpha} = \frac{1}{\beta(\alpha-1)}.$$

(b) $\pi(\theta|\mathbf{x}) \propto \theta^n e^{-\theta \Sigma x_j} \theta^{\alpha-1} e^{-\theta/\beta} = \theta^{n+\alpha-1} e^{-\theta(\Sigma x_j + \beta^{-1})}$ which is a gamma density. Therefore the actual posterior distribution is

$$\pi(\theta|\mathbf{x}) = \frac{\theta^{n+\alpha-1} e^{-\theta(\Sigma x_j + \beta^{-1})}}{\Gamma(n+\alpha)(\Sigma x_j + \beta^{-1})^{-n-\alpha}}$$

with mean

$$E(\theta|\mathbf{x}) = \frac{n+\alpha}{\Sigma x_j + \beta^{-1}}.$$

5.9 (a)

$$
\begin{aligned}
f(x) &= \int f(x|\theta)b(\theta)d\theta \\
&= \binom{r+x-1}{x} \frac{\Gamma(a+b)}{\Gamma(a)\Gamma(b)} \int_0^1 \theta^{r+a-1}(1-\theta)^{b+x-1}d\theta \\
&= \binom{r+x-1}{x} \frac{\Gamma(a+b)}{\Gamma(a)\Gamma(b)} \frac{\Gamma(r+a)\Gamma(b+x)}{\Gamma(r+a+b+x)} \\
&= \frac{\Gamma(r+x)}{\Gamma(r)x!} \frac{\Gamma(a+b)}{\Gamma(a)\Gamma(b)} \frac{\Gamma(a+r)\Gamma(b+x)}{\Gamma(a+r+b+x)}
\end{aligned}
$$

(b)

$$
\begin{aligned}
\pi(\theta|\mathbf{x}) &\propto \prod f(x_j|\theta)b(\theta) \\
&\propto \theta^{nr}(1-\theta)^{\Sigma x_j} \theta^{a-1}(1-\theta)^{b-1} \\
&= \theta^{a+nr-1}(1-\theta)^{b+\Sigma x_j-1}
\end{aligned}
$$

Hence, $\pi(\theta|\mathbf{x})$ is Beta with parameters

$$
\begin{aligned}
a^* &= \alpha + nr \\
b^* &= b + \sum x_j \\
E(\theta|\mathbf{x}) &= \frac{a^*}{a^* + b^*} = \frac{a+nr}{a+nr+b+\sum x_j}
\end{aligned}
$$

5.10 (a)

$$
\begin{aligned}
f(x) &= \int f(x|\theta)b(\theta)d\theta = \int_0^\infty \sqrt{\frac{\theta}{2\pi}} e^{-\frac{\theta}{2}(x-\mu)^2} \frac{\beta^\alpha}{\Gamma(\alpha)} \theta^{\alpha-1} e^{-\beta\theta} d\theta \\
&= \frac{\beta^\alpha}{(2\pi)^{1/2}\Gamma(\alpha)} \int_0^\infty \theta^{\frac{1}{2}+\alpha-1} e \exp\left\{-\theta\left[\frac{1}{2}(x-\mu)^2+\beta\right]\right\} d\theta \\
&= \frac{\beta^\alpha}{(2\pi)^{1/2}\Gamma(\alpha)} \frac{\Gamma\left(\alpha+\frac{1}{2}\right)}{\left[\frac{1}{2}(x-\mu)^2+\beta\right]^{a+\frac{1}{2}}} \\
&= \frac{\Gamma\left(\alpha+\frac{1}{2}\right)}{\sqrt{2\pi}\beta\Gamma(\alpha)} \left[1+\frac{1}{2\beta}(x-\mu)^2\right]^{-\alpha-\frac{1}{2}}
\end{aligned}
$$

(b)
$$\pi\left(\theta|\mathbf{x}\right) \propto \left[\prod f(x_j|\theta)\right]\pi(\theta)$$

$$\propto \theta^{n/2}\exp\left[-\frac{\theta}{2}\sum(x_j-\mu)^2\right]\theta^{\alpha-1}e^{-\beta\theta}$$

$$= \theta^{\frac{n}{2}+\alpha-1}\exp\left\{-\theta\left[\frac{1}{2}\sum(x_j-\mu)^2+\beta\right]\right\} = \theta^{\alpha^*-1}e^{-\theta\beta^*}$$

where $\alpha^* = \alpha+\frac{n}{2}$, $\beta^* = \beta+\frac{1}{2}\sum(x_j-\mu)^2$. Therefore, $b(\theta|\mathbf{x})$ is gamma (α^*,β^*) and $E(\theta|\mathbf{x}) = \frac{\alpha^*}{\beta^*} = \frac{\alpha+\frac{n}{2}}{\beta+\frac{1}{2}\sum(x_j-\mu)^2}$

5.11 (a)
$$f\left(x|\theta_1,\theta_2\right) = \sqrt{\frac{\theta_2}{2\pi}}\exp\left[-\frac{\theta_2}{2}\left(x-\theta_1\right)^2\right]$$

$$b\left(\theta_1|\theta_2\right) = \sqrt{\frac{\theta_2}{2\pi\sigma^2}}\exp\left[-\frac{\theta_2}{2\sigma^2}\left(\theta_1-\mu\right)^2\right]$$

$$b\left(\theta_2\right) = \frac{\beta^\alpha}{\Gamma(\alpha)}\theta_2^{\alpha-1}e^{-\beta\theta_2}$$

$$\pi\left(\theta_1,\theta_2|\mathbf{x}\right) \propto \left[\prod_{j=1}^{r}f(x_j|\theta_1,\theta_2)\right]\pi\left(\theta_1|\theta_2\right)\pi\left(\theta_2\right)$$

$$\propto \theta_2^{n/2}\exp\left[-\frac{\theta_2}{2}\sum(x_j-\theta_1)^2\right]\theta_2^{1/2}$$

$$\times\exp\left[-\frac{\theta_2}{2\sigma_2}\left(\theta_1-\mu\right)^2\right]\theta_2^{\alpha-1}e^{-\beta\theta_2}$$

$$= \theta_2^{\alpha+\frac{n+1}{2}-1}$$

$$\times\exp\left\{-\theta_2\left[\beta+\frac{1}{2}\left(\frac{\theta_1-\mu}{\sigma}\right)^2+\frac{1}{2}\sum(x_j-\theta_1)^2\right]\right\}$$

$$\pi\left(\theta_1|\theta_2,\mathbf{x}\right) \propto \pi\left(\theta_1,\theta_2|\mathbf{x}\right)$$

$$\propto \exp\left[-\frac{\theta_2}{2}\left(\frac{\theta_1^2}{\sigma^2}-\frac{2\mu\theta_1}{\sigma^2}+n\theta_1^2-2\theta_1\sum x_j\right)\right]$$

$$= \exp-\frac{1}{2}\left[\theta_1^2\left(\frac{\theta_2}{\sigma_2}+n\theta_2\right)-2\theta_1\left(\frac{\mu\theta_2}{\sigma^2}+\theta_2\sum x_j\right)\right]$$

which is normal with variance $\sigma_*^2 = \left[\frac{\theta_2}{\sigma^2}+n\theta_2\right]^{-1} = \frac{\sigma^2}{\theta_2(1+n\sigma^2)}$ and mean μ_* which satisfies $\frac{\mu_*}{\sigma_*^2} = \frac{\mu\theta_2}{\sigma^2}+\theta_2 n\bar{x}$. Then, $\mu_* = \frac{\mu}{1+n\sigma^2}+\frac{n\sigma^2\bar{x}}{1+n\sigma^2}$.
For the posterior distribution of θ_2,

$$\pi(\theta_2|\mathbf{x}) \quad \propto \quad \int \pi(\theta_1, \theta_2|\mathbf{x}) \, d\theta_1$$

$$= \quad \theta_2^{\alpha + \frac{n+1}{2} - 1} e^{-\theta_2 \beta}$$

$$\times \int \exp\left\{-\frac{\theta_2}{2}\left[\left(\frac{\theta_1 - \mu}{\sigma}\right)^2 + \sum(x_j - \theta_1)^2\right]\right\} d\theta_1$$

Now $\sum(x_j - \theta_1)^2 = \sum(x_j - \overline{x})^2 + n(\overline{x} - \theta_1)^2$ and therefore

$$\pi(\theta_2|\mathbf{x}) \quad \propto \quad \theta_2^{\alpha + \frac{n+1}{2} - 1} \exp\left\{-\theta_2\left[\beta + \frac{1}{2}\sum(x_j - \overline{x})^2\right]\right\}$$

$$\times \int \exp\left\{-\frac{\theta_2}{2}\left[\left(\frac{\theta_1 - \mu}{\sigma}\right)^2 + n(\overline{x} - \theta_1)^2\right]\right\} d\theta_1$$

In order to evaluate the integral, complete the square as follows

$$-\frac{\theta_2}{2}\left[\left(\frac{\theta_1 - \mu}{\sigma}\right)^2 + n(\overline{x} - \theta_1)^2\right]$$

$$= -\frac{\theta_2}{2}\left[\theta_1^2(1/\sigma^2 + n) - 2\theta_1(\mu/\sigma^2 + n\overline{x}) + \mu^2/\sigma^2 + n\overline{x}^2\right]$$

$$= -\frac{1}{2}\theta_2(1/\sigma^2 + n)\left(\theta_1 - \frac{\mu + n\sigma^2\overline{x}}{1 + n\sigma^2}\right)^2 + \frac{\theta_2}{2}\left[\frac{(\mu + n\sigma^2\overline{x})^2}{\sigma^2(1 + n\sigma^2)} - \frac{\mu^2}{\sigma^2} - n\overline{x}\right]$$

The first term is a normal density and integrates to $[\theta_2(1/\sigma^2 + n)]^{-1/2}$. The second term does not involve θ_1 and so factors out of the integral. The posterior density contains θ_2 raised to the $\alpha + \frac{n}{2} - 1$ power and an exponential term involving θ_2 multiplied by

$$\beta + \frac{1}{2}\sum(x_j - \overline{x})^2 + \frac{1}{2}\left[\frac{(\mu + n\sigma^2\overline{x})^2}{\sigma^2(1 + n\sigma^2)} - \frac{\mu^2}{\sigma^2} - n\overline{x}\right]$$

$$= \beta + \frac{1}{2}\sum(x_j - \overline{x})^2 + \frac{n(\overline{x} - \mu)^2}{2(1 + n\sigma^2)} .$$

This constitutes a gamma density with the desired parameters.

(b) Because the mean of Θ_1 given Θ_2 and $\mathbf{x}$ does not depend on θ_2 it is also the mean of Θ_1 given just $\mathbf{x}$ which is μ_*. The mean of Θ_2 given $\mathbf{x}$ is the ratio of the parameters or

$$\frac{\alpha + n/2}{\beta + \frac{1}{2}\sum(x_j - \overline{x})^2 + \frac{n(\overline{x} - \mu)^2}{2(1 + n\sigma^2)}} .$$

5.12 $f(x; p) = \binom{n}{x}p^x(1 - p)^{n-x} = \binom{n}{x}(1 - p)^n\left(\frac{p}{1 - p}\right)^x$

$$= \binom{n}{x}(1 - p)^n \exp\left\{x \log[p/(1 - p)]\right\}$$

Save 54% Liberal Discount

Subscribe to the new American Prospect

☐ **Save 54%** Get 1 year / 22 issues / just $29.95

☐ **Save 69%** Get 3 years / 66 issues / just $59.95

NAME

ADDRESS

CITY   STATE   ZIP   APT #

E-MAIL ADDRESS (Optional)

Payment Method

☐ Check
(to "The American Prospect" enclosed)

Charge my ☐ Mastercard ☐ Visa

B0815I

BUSINESS REPLY MAIL

FIRST-CLASS MAIL   PERMIT NO. 121   MT MORRIS, IL

POSTAGE WILL BE PAID BY ADDRESSEE

THE AMERICAN
P R O S P E C T

PO BOX 601
MT MORRIS, IL 61054-7531

If we let $p(x) = \binom{n}{x}$, $q(\theta) = (1 - p)^{-n}$, and $\theta = -\log\frac{p}{1-p}$, then $p = \frac{1}{1+e^\theta}$ and $q(\theta) = (1 - p)^{-n} = \left(\frac{e^\theta}{1+e^\theta}\right)^{-n}$.

5.13 $f(x) = \frac{\Gamma(\alpha+x)}{\Gamma(\alpha)x!}\frac{1}{(\frac{1+\beta}{\beta})^\alpha}e^{\log(\frac{1}{1+\beta})}$. Then let $\theta = \log(1 + \beta)$. Then $\frac{1+\beta}{\beta} = \frac{1}{1-e^{-\theta}}$. The density function can be written $f(x) = \frac{\Gamma(\alpha+x)}{\Gamma(\alpha)x!}\frac{1}{\left(\frac{1}{1-e^{-\theta}}\right)^\alpha}e^{-x\theta}$. This is of the desired form with $p(x) = \frac{\Gamma(\alpha+x)}{\Gamma(\alpha)x!}$ and $q(\theta) = \left(\frac{1}{1-e^{-\theta}}\right)^\alpha$.

5.14
$$L(\theta) = \prod_{j=1}^{n} f(x_j; \theta) = \prod_{j=1}^{n} \frac{p(x_j)e^{-\theta x_j}}{q(\theta)}$$

$$\ell(\theta) = \sum_{j=1}^{n} \log p(x_j) - \theta \sum_{j=1}^{n} x_j - n \log q(\theta)$$

$$\ell'(\theta) = -\sum_{j=1}^{n} x_j - n\frac{q'(\theta)}{q(\theta)}$$

$$\text{Therefore,} \quad -\frac{q'(\hat\theta)}{q(\hat\theta)} = \bar{x}$$

$$\text{But,} \quad -\frac{q'(\theta)}{q(\theta)} = E(x) = \mu(\theta)$$

$$\text{and so,} \quad \mu(\hat\theta) = \bar{x}$$

5.15 For the mean,
$$\log f(x; \theta) = \log p(m, x) - m\theta x - m \log q(\theta)$$

$$\frac{\partial \log f(x; \theta)}{\partial \theta} = \frac{\partial f(x; \theta)}{\partial \theta}\frac{1}{f(x; \theta)} = \left[-mx - \frac{mq'(\theta)}{q(\theta)}\right]$$

$$\frac{\partial f(x; \theta)}{\partial \theta} = \left[-mx - \frac{mq'(\theta)}{q(\theta)}\right]f(x; \theta)$$

$$0 = \int\frac{\partial f(x; \theta)}{\partial \theta}dx = -m\int xf(x; \theta)dx - \frac{mq'(\theta)}{q(\theta)}\int f(x; \theta)dx$$

$$= -mE(X) - \frac{mq'(\theta)}{q(\theta)}$$

$$E(X) = q'(\theta)/q(\theta) = \mu(\theta)$$

For the variance,
$$\frac{\partial f(x; \theta)}{\partial \theta} = -m[x - \mu(\theta)]f(x; \theta)$$

$$\frac{\partial^2 f(x;\theta)}{\partial\theta^2} = m\mu'(\theta)f(x;\theta) - m[x - \mu(\theta)]\frac{\partial f(x;\theta)}{\partial\theta}$$

$$= m\mu'(\theta)f(x;\theta) + m^2[x - \mu(\theta)]^2 f(x;\theta)$$

$$0 = \int \frac{\partial^2 f(x;\theta)}{\partial\theta^2}dx = m\mu'(\theta)\int f(x;\theta)dx + m^2\int[x - \mu(\theta)]^2 f(x;\theta)dx$$

$$= m\mu'(\theta) + m^2 Var(X)$$

$$Var(X) = -\mu'(\theta)/m$$

5.16 (a)

$$f_X(x) = \int f_{X|\Theta}(x|\theta)\pi(\theta)d\theta = \int \frac{p(x)e^{-\theta x}}{q(\theta)}\frac{[q(\theta)]^{-k}e^{\theta\mu k}}{c(\mu,k)}d\theta$$

$$= \frac{p(x)}{c(\mu,k)}\int [q(\theta)]^{-(k+1)}\exp\left[-\theta\frac{(x+\mu k)(k+1)}{k+1}\right]d\theta$$

$$= \frac{p(x)}{c(\mu,k)}c\left(\frac{x+\mu k}{k+1},k+1\right)$$

(b)

$$\pi(\theta|\mathbf{x}) \propto \frac{\left[\prod_{j=1}^{n}p(x_j)\right]e^{-\theta\Sigma x_j}}{q(\theta)^n}\frac{[q(\theta)]^{-k}e^{-\theta\mu k}}{c(\mu,k)}$$

$$\propto [q(\theta)]^{-(k+n)}\exp\left(-\theta\frac{\mu k+\sum x_j}{k+n}\right)(k+n)$$

$$\propto [q(\theta)]^{-k^*}\exp[-\theta\mu^*(k+n)]$$

which is of the same form as $\pi(\theta)$ with parameters

$$k^* = k+n$$

$$\mu^* = \frac{\mu k+\sum x_j}{k+n} = \frac{k}{k+n}\mu + \frac{n}{k+n}\bar{x}$$

5.17 (a)

$$E(X) = E[E(X|\Theta_1,\Theta_2)] = E(\Theta_1)$$

$$Var(X) = E[Var(X|\Theta_1,\Theta_2)] + Var[E(X|\Theta_1,\Theta_2)]$$

$$= E(\Theta_2) + Var(\Theta_1)$$

(b) $f_{X|\Theta_1,\Theta_2}(x|\theta_1,\theta_2) = (2\pi\theta_2^2)^{-1/2}\exp\left[-\frac{1}{2\theta_2}(x-\theta_1)^2\right]$

$$f_X(x) = \int\int(2\pi\theta_2^2)^{-1/2}\exp\left[-\frac{1}{2\theta_2}(x-\theta_1)^2\right]\pi(\theta_1,\theta_2)d\theta_1 d\theta_2$$

$$= \int\int(2\pi\theta_2^2)^{-1/2}\exp\left[-\frac{1}{2\theta_2}(x-\theta_1)^2\right]\pi_1(\theta_1)\pi_2(\theta_2)d\theta_1 d\theta_2.$$

We are also given, $f_{Y|\Theta_2}(y|\theta_2) = (2\pi\theta_2^2)^{-1/2}\exp\left(-\frac{1}{2\theta_2}y^2\right)$. Let $Z = Y + \Theta_1$.
Then

$$f_{Z,|\Theta_2}(z|\theta_2) = \int f_{Y|\theta_2}(z - \theta_1|\theta_2)\,\pi_1(\theta_1)d\theta_1$$

and

$$
\begin{aligned}
f_Z(z) &= \int f_{Z,|\Theta_2}(z|\theta_2)\,\pi_2(\theta_2)d\theta_2 \\
&= \int\int f_{Y|\theta_2}(z - \theta_1|\theta_2)\,\pi_1(\theta_1)d\theta_1\pi_2(\theta_2)d\theta_2 \\
&= \int\int (2\pi\theta_2^2)^{-1/2}\exp\left[-\frac{1}{2\theta_2}(z - \theta_1)^2\right]\pi_1(\theta_1)d\theta_1\pi_2(\theta_2)d\theta_2 \\
&= f_X(x).
\end{aligned}
$$

5.18
$$
\begin{aligned}
f_X(x) &= \int \frac{e^{-\theta}\theta^x}{x!}\pi_1(\theta)d\theta = \int \frac{e^{-\theta}\theta^x}{x!}\pi(\theta - x)d\theta \\
f_Y(y) &= \frac{e^{-\alpha}\alpha^y}{y!} \\
f_Z(z) &= \int \frac{e^{-\theta}\theta^z}{z!}\pi(\theta)d\theta
\end{aligned}
$$

Let $W = Y + Z$. Then

$$
\begin{aligned}
f_W(w) &= \sum_{y=0}^{w} f_Y(y)f_Z(w - y) \\
&= \sum_{y=0}^{w} \frac{e^{-\alpha}\alpha^y}{y!}\int \frac{e^{-\theta}\theta^{(w-y)}}{(w - y)!}\pi(\theta)d\theta \\
&= \int \frac{e^{-(\alpha+\theta)}(\alpha + \theta)^w}{w!}\sum_{y=0}^{w}\binom{w}{y}\left(\frac{\alpha}{\alpha + \theta}\right)^y\left(\frac{\theta}{\alpha + \theta}\right)^{w-y}\pi(\theta)d\theta \\
&= \int \frac{e^{-(\alpha+\theta)}(\alpha + \theta)^w}{w!}\pi(\theta)d\theta
\end{aligned}
$$

with the last line following because the sum contains binomial probabilities.
Let $r = \alpha + \theta$ and so

$$
\begin{aligned}
f_W(w) &= \int \frac{e^{-r}r^w}{w!}\pi(r - \alpha)dr \\
&= f_X(x).
\end{aligned}
$$

5.19 (a)
$$f(x|\theta) = \frac{p(x)e^{-\theta x}}{q(\theta)}$$

$$M_X(z|\theta) = \int \frac{p(x)e^{-\theta x}}{q(\theta)} e^{zx} dx$$

$$= \frac{1}{q(\theta)} \int p(x)e^{-(\theta-z)x} dx$$

$$= \frac{q(\theta-z)}{q(\theta)}.$$

Hence, $M_S(z) = [q(\theta-z)/q(\theta)]^n = q_*(\theta-z)/q_*(\theta)$ where $q_*(\theta) = [q(\theta)]^n$.

The distribution of S is of same form as X except that $p_*(x)$ may depend on n. Hence

$$f_{S|\Theta}(s|\theta) = p_n(s)\frac{e^{-\theta s}}{q(\theta)^n}$$

(b)
$$\pi(\theta|\mathbf{x}) \propto f(\mathbf{x}, \theta) = f(\mathbf{x}|\theta)\pi(\theta)$$

$$= \prod f(x_j|\theta)\pi(\theta)$$

$$\propto \frac{e^{-\theta\Sigma x_j}}{[q(\theta)]^n}\pi(\theta)$$

$$= \frac{e^{-\theta s}}{[q(\theta)]^n}\pi(\theta)$$

But $\pi(\theta|S) \propto f(s|\theta)\pi(\theta) \propto \frac{e^{-\theta t}}{[q(\theta)]}\pi(\theta)$.

5.20 (a) Let N be Poisson(λ).

$$f(x) = \sum_{n=x}^{\infty} \frac{n!}{x!(n-x)!}p^x(1-p)^{n-x}e^{-\lambda}\lambda^n/n!$$

$$= \left(\frac{p}{1-p}\right)^x \frac{e^{-\lambda}}{x!} \sum_{n=x}^{\infty} \frac{[(1-p)\lambda]^n}{(n-x)!}$$

$$= \left(\frac{p}{1-p}\right)^x \frac{e^{-\lambda}}{x!} \sum_{n=0}^{\infty} \frac{[(1-p)\lambda]^{n+x}}{n!}$$

$$= \left(\frac{p}{1-p}\right)^x \frac{e^{-\lambda}}{x!}[(1-p)\lambda]^x e^{(1-p)\lambda}$$

$$= e^{-p\lambda}(p\lambda)^x/x!, \text{ a Poisson distribution with parameter } p\lambda.$$

(b) Let N be binomial(m, r).

$$f(x) = \sum_{n=x}^{m} \frac{n!}{x!(n-x)!}p^x(1-p)^{n-x}\frac{m!}{n!(m-n)!}r^n(1-r)^{m-n}$$

$$= \left(\frac{p}{1-p}\right)^x (1-r)^m\frac{m!}{x!}\sum_{n=x}^{m}\left(\frac{[1-p]r}{1-r}\right)^n \frac{1}{(n-x)!(m-n)!}$$

$$= \left(\frac{p}{1-p}\right)^x (1-r)^m \frac{m!}{x!} \sum_{n=0}^{m-x} \left(\frac{[1-p]r}{1-r}\right)^{n+x} \frac{1}{n!(m-n-x)!}$$

$$= \left(\frac{p}{1-p}\right)^x (1-r)^m \frac{m!}{x!} \left(\frac{[1-p]r}{1-r}\right)^x \frac{1}{(m-x)!}$$

$$\times \sum_{n=0}^{m-x} \left(\frac{[1-p]r}{1-rp}\right)^n \left(\frac{1-r}{1-rp}\right)^{-n} \frac{(m-x)!}{n!(m-n-x)!}$$

$$= \left(\frac{pr}{1-r}\right)^x (1-r)^m \frac{m!}{x!(m-x)!} \left(\frac{1-r}{1-rp}\right)^{-m+x}$$

$$\times \sum_{n=0}^{m-x} \left(\frac{[1-p]r}{1-rp}\right)^n \left(\frac{1-r}{1-rp}\right)^{m-n-x} \frac{(m-x)!}{n!(m-n-x)!}$$

$$= \frac{m!}{x!(m-x)!} (pr)^x (1-rp)^{m-x}, \text{ which is binomial } (m, pr).$$

(c) Let N be negative binomial (r, β).

$$f(x) = \sum_{n=x}^{\infty} \frac{n!}{x!(n-x)!} p^x (1-p)^{n-x}$$

$$\times \left(\frac{\beta}{1+\beta}\right)^n \left(\frac{1}{1+\beta}\right)^r \frac{r(r+1)\cdots(r+n-1)}{n!}$$

$$= \left(\frac{p}{1-p}\right)^x \left(\frac{1}{1+\beta}\right)^r \frac{1}{x!} \sum_{n=x}^{\infty} \left[\frac{\beta(1-p)}{1+\beta}\right]^n \frac{r(r+1)\cdots(r+n-1)}{(n-x)!}$$

$$= \left(\frac{p}{1-p}\right)^x \left(\frac{1}{1+\beta}\right)^r \frac{1}{x!} \sum_{n=0}^{\infty} \left[\frac{\beta(1-p)}{1+\beta}\right]^{n+x}$$

$$\times \frac{r(r+1)\cdots(r+n+x-1)}{n!}$$

$$= \left(\frac{p\beta}{1+\beta}\right)^x \left(\frac{1}{1+\beta}\right)^r \frac{r(r+1)\cdots(r+x-1)}{x!}$$

$$\times \sum_{n=0}^{\infty} \left[\frac{\beta(1-p)}{1+\beta}\right]^n \frac{(r+x)\cdots(r+n+x-1)}{n!}$$

and the summand is almost a negative binomial density with the term

$$\left[1 - \frac{\beta(1-p)}{1+\beta}\right]^{r+x} = \left(\frac{1+p\beta}{1+\beta}\right)^{r+x}$$

missing. Place it in the sum so the sum is one and then divide by it to produce

$$f(x) = \left(\frac{1+p\beta}{1+\beta}\right)^{-r-x}$$

$$= \left(\frac{p\beta}{1+p\beta}\right)^x \left(\frac{1}{1+p\beta}\right)^r \frac{r(r+1)\cdots(r+x-1)}{x!}$$

which is negative binomial with parameters r and $p\beta$.

5.21 Let D be the die. Then

$$\Pr(D = 2|2, 3, 4, 1, 4)$$

$$= \frac{\Pr(2, 3, 4, 1, 4|D = 2)\Pr(D = 2)}{\Pr(2, 3, 4, 1, 4|D = 1)\Pr(D = 1) + \Pr(2, 3, 4, 1, 4|D = 2)\Pr(D = 2)}$$

$$= \frac{\frac{1}{6}\frac{1}{6}\frac{3}{6}\frac{1}{6}\frac{3}{6}\frac{1}{2}}{\frac{3}{6}\frac{1}{6}\frac{1}{6}\frac{1}{6}\frac{1}{6}\frac{1}{2} + \frac{1}{6}\frac{1}{6}\frac{3}{6}\frac{1}{6}\frac{3}{6}\frac{1}{2}} = \frac{3}{4}.$$

5.22 (a) $\Pr(Y = 0) = \int_0^1 e^{-\theta}(1)d\theta = 1 - e^{-1} = 0.63212 > 0.35.$
(b) $\Pr(Y = 0) = \int_0^1 \theta^2(1)d\theta = 1/3 < 0.35.$
(c) $\Pr(Y = 0) = \int_0^1 (1 - \theta)^2(1)d\theta = 2/3 > 0.35.$
Only (a) and (c) are possible.

5.23 $\Pr(H = 1/4|d = 1)$

$$= \frac{\Pr(d = 1|H = 1/4)\Pr(H = 1/4)}{\Pr(d = 1|H = 1/4)\Pr(H = 1/4) + \Pr(d = 1|H = 1/2)\Pr(H = 1/2)}$$

$$= \frac{\frac{1}{4}\frac{4}{5}}{\frac{1}{4}\frac{4}{5} + \frac{1}{2}\frac{1}{5}} = \frac{2}{3}.$$

5.24 The posterior *pdf* is proportional to

$$\frac{e^{-\theta}\theta^0}{0!}e^{-\theta} = e^{-2\theta}.$$

This is an exponential distribution. The *pdf* is $\pi(\theta|y = 0) = 2e^{-2\theta}.$

5.25 The posterior *pdf* is proportional to

$$\frac{e^{-\theta}\theta^1}{1!}\theta e^{-\theta} = \theta^2 e^{-2\theta}.$$

This is a gamma distribution with parameters 3 and 0.5. The *pdf* is $\pi(\theta|y = 1) = 4\theta^2 e^{-2\theta}.$

5.26 From Example 5.14 the posterior distribution is gamma with $\alpha_* = 50 + 177 = 227$ and $\theta_* = 0.002/[1{,}850(0.002) + 1] = 1/2{,}350$. The mean is $\alpha_*\theta_* = 0.096596$ and the variance is $\alpha_*\theta_*^2 = 0.000041105$. The coefficient of variation is $\sqrt{\alpha_*\theta_*^2}/(\alpha_*\theta_*) = 1/\sqrt{\alpha_*} = 0.066372.$

5.27 The posterior *pdf* is proportional to

$$\binom{3}{1}\theta^1(1-\theta)^{3-1}6\theta(1-\theta) \propto \theta^2(1-\theta)^3.$$

This is a beta distribution with *pdf* $\pi(\theta|r=1) = 60\theta^2(1-\theta)^3$.

5.28 The equations are $\alpha\beta = 0.14$ and $\alpha\beta^2 = 0.0004$. The solution is $\alpha = 49$ and $\beta = 1/350$. From Example 5.14 the posterior distribution is gamma with $\alpha_* = 49 + 110 = 159$ and $\beta_* = (1/350)/[620(1/350) + 1] = 1/970$. The mean is $159/970 = 0.16392$ and the variance is $159/970^2 = 0.00016899$.

5.29 The prior exponential distribution is also a gamma distribution with $\alpha = 1$ and $\beta = 2$. From Example 5.14 the posterior distribution is gamma with $\alpha_* = 1 + 3 = 4$ and $\beta_* = 2/[1(2) + 1] = 2/3$. The *pdf* is $\pi(\lambda|y = 3) = 27\lambda^3 e^{-3\lambda/2}/32$.

5.30 (a) The posterior distribution is proportional to

$$\binom{3}{2}\theta^2(1-\theta)280\theta^3(1-\theta)^4 \propto \theta^5(1-\theta)^5$$

which is a beta distribution. The *pdf* is $\pi(\theta|y = 2) = 2772\theta^5(1-\theta)^5$.
(b) The mean is $6/(6+6) = 0.5$.

5.31 The posterior *pdf* is proportional to

$$te^{-t5}te^{-t} = t^2e^{-t6}$$

which is a gamma distribution with $\alpha = 3$ and $\beta = 1/6$. The posterior *pdf* is $\pi(t|x = 5) = 108t^2e^{-6t}$.

4.2   SECTION 5.3

5.32
$$\lambda_0 = (1.96/0.05)^2 = 1{,}536.64$$
$$E(X) = \int_0^{100} \frac{100x - x^2}{5{,}000}\,dx = 33\frac{1}{3}$$
$$E(X^2) = \int_0^{100} \frac{100x^2 - x^3}{5{,}000}\,dx = 1{,}666\frac{2}{3}$$
$$Var(X) = 1{,}666\frac{2}{3} - (33\frac{1}{3})^2 = 555\frac{5}{9}$$
$$n\lambda = 1{,}536.64\left[1 + \left(\frac{\sqrt{555\frac{5}{9}}}{33\frac{1}{3}}\right)^2\right] = 2{,}304.96$$

2,305 claims are needed.

5.33 $0.81 = n\lambda/\lambda_0$. $0.64 = n\lambda \left[\lambda_0 \left(1 + \frac{\alpha\beta^2}{\alpha^2\beta^2} \right) \right]^{-1} = 0.81(1 + \alpha^{-1})^{-1}$, $\alpha = 3.7647$. $\alpha\beta = 100$, $\beta = 26.5625$.

5.34 $\lambda_0 = 1,082.41$. $\mu = 600$, Estimate the variance as $\sigma^2 = \frac{0^2 + 75^2 + (-75)^2}{2} = 75^2$. The standard for full credibility is $1,082.41(75/600)^2 = 16.913$. $Z = \sqrt{3/16.913} = 0.4212$. The credibility pure premium is

$$0.4212(475) + 0.5788(600) = 547.35.$$

5.35 $\mu = s\beta\theta_Y$, $\sigma^2 = s\beta\sigma_Y^2 + s\beta(1 + \beta)\theta_Y^2$ where s is used in place of the customary negative binomial parameter, r. Then

$$1.645 = \frac{r\mu\sqrt{n}}{\sigma} = \frac{0.05\sqrt{n}s\beta\theta_Y}{\sqrt{s\beta\sigma_Y^2 + s\beta(1 + \beta)_Y^2}}$$

and so

$$1,082.41 = \frac{ns^2\beta^2\theta_Y^2}{s\beta\sigma_Y^2 + s\beta(1 + \beta)\theta_Y^2} = ns\beta \left(\frac{\theta_Y^2}{\sigma_Y^2 + \theta_Y^2(1 + \beta)} \right).$$

The standard for full credibility is

$$ns\beta = 1,082.41 \left(1 + \beta + \frac{\sigma_Y^2}{\theta_Y^2} \right).$$

Partial credibility is obtained by taking the square root of ratio of the number of claims to the standard for full credibility.

5.36 $\lambda_0 = (2.2414/.03)^2 = 5,582.08$. 5,583 claims are required.

5.37
$$E(X) = \int_0^{200,000} \frac{x}{200,000} dx = 100,000$$

$$E(X^2) = \int_0^{200,000} \frac{x^2}{200,000} dx = 13,333,333,333\frac{1}{3}$$

$$Var(X) = 3,333,333,333\frac{1}{3}$$

The standard for full credibility is

$$1,082.41(1 + 3,333,333,333.33/10,000,000,000) = 1,443.21$$

$Z = \sqrt{1{,}082/1{,}443.21} = 0.86586.$

5.38 $(1.645/0.06)^2(1 + 7{,}500^2/1{,}500^2) = 19{,}543.51$ or $19{,}544$ claims.

5.39 $Z = \sqrt{6{,}000/19{,}543.51} = 0.55408.$ The credibility estimate is

$$0.55408(15{,}600{,}000) + 0.44592(16{,}500{,}000) = 16{,}001{,}328.$$

5.40 For the standard for estimating the number of claims, $800 = (y_p/0.05)^2$ and so $y_p = \sqrt{2}.$

$$
\begin{aligned}
E(X) &= \int_0^{100} 0.0002x(100 - x)dx = 33\frac{1}{3}, \\
E(X^2) &= \int_0^{100} 0.0002x^2(100 - x)dx = 1{,}666\frac{2}{3}, \\
Var(X) &= 1{,}666\frac{2}{3} - (33\frac{1}{3})^2 = 555\frac{5}{9}.
\end{aligned}
$$

The standard for full credibility is $(\sqrt{2}/0.1)^2[1 + 555\frac{5}{9}/(33\frac{1}{3})^2] = 300$

5.41 $1{,}000 = (1.96/0.1)^2(1 + c^2).$ The solution is the coefficient of variation, $1.2661.$

5.42 $Z = \sqrt{10{,}000/17{,}500} = 0.75593.$ The credibility estimate is

$$0.75593(25{,}000{,}000) + 0.24407(20{,}000{,}000) = 23{,}779{,}650.$$

5.43 The standard for estimating claim numbers is $(2.326/0.05)^2 = 2{,}164.11.$ For estimating the amount of claims we have $E(X) = \int_1^\infty 2.5x^{-2.5}dx = 5/3,$ $E(X^2) = \int_1^\infty 2.5x^{-1.5}dx = 5,$ and $Var(X) = 5 - (5/3)^2 = 20/9.$ Then $2164.11 = (1.96/K)^2[1 + (20/9)/(25/9)],$ $K = 0.056527$

5.44 $E(X) = 0.5(1) + 0.3(2) + 0.2(10) = 3.1,$ $E(X^2) = 0.5(1) + 0.3(4) + 0.2(100) = 21.7,$ $Var(X) = 21.7 - 3.1^2 = 12.09.$ The standard for full credibility is $(1.645/.1)^2(1 + 12.09/3.1^2) = 611.04$ and so 612 claims are needed.

5.45 $3415 = (1.96/k)^2(1 + 4),$ $k = 0.075$ or $7.5\%.$

5.46 $Z = \sqrt{n/F},$ $R = ZO + (1 - Z)P,$ $Z = (R - P)/(O - P) = \sqrt{n/F}.$ $n/F = (R - P)^2/(O - P)^2.$ $n = \frac{F(R-P)^2}{(O-P)^2}.$

4.3   SECTION 5.4

5.47 (a) $\pi(\theta_{ij}) = 1/6$ for die i and spinner j.
(b)(c) The calculations are in Table 4.1.

Table 4.1   Calculations for Exercise 5.47(b) and (c)

i	j	$\Pr(X = 0\vert\theta_{ij})$	$\Pr(X = 3\vert\theta_{ij})$	$\Pr(X = 8\vert\theta_{ij})$	$\mu(\theta_{ij})$	$v(\theta_{ij})$
1	1	25/30	1/30	4/30	35/30	6,725/900
1	2	25/30	2/30	3/30	30/30	5,400/900
1	3	25/30	4/30	1/30	20/30	2,600/900
2	1	10/30	4/30	16/30	140/30	12,200/900
2	2	10/30	8/30	12/30	120/30	10,800/900
2	3	10/30	16/30	4/30	80/30	5,600/900

(d) $\Pr(X_1 = 3) = (1 + 2 + 4 + 4 + 8 + 16)/(30 \cdot 6) = 35/180$.
(e) The calculations are in Table 4.2.

Table 4.2   Calculations for Exercise 5.47(e)

i	j	$\Pr(X_1 = 3\vert\theta_{ij})$	$\Pr(\Theta = \theta_{ij}\vert X_1 = 3) = \frac{\Pr(X_1=3\vert\theta_{ij})(1/6)}{35/180}$
1	1	1/30	1/35
1	2	2/30	2/35
1	3	4/30	4/35
2	1	4/30	4/35
2	2	8/30	8/35
2	3	16/30	16/35

(f) The calculations are in Table 4.3.

Table 4.3   Calculations for Exercise 5.47(f)

x_2	$\Pr(X_2 = x_2\vert X_1 = 3) = \sum \Pr(X_2 = x_2\vert\Theta = \theta_{ij})\Pr(\Theta = \theta_{ij}\vert X_1 = 3)$
0	$\frac{1}{30}\frac{1}{35}[25(1) + 25(2) + 25(4) + 10(4) + 10(8) + 10(16)] = 455/1{,}050$
3	$\frac{1}{30}\frac{1}{35}[1(1) + 2(2) + 4(4) + 4(4) + 8(8) + 16(16)] = 357/1{,}050$
8	$\frac{1}{30}\frac{1}{35}[4(1) + 3(2) + 1(4) + 16(4) + 12(8) + 4(16)] = 238/1{,}050$

(g)

$$
\begin{aligned}
E(X_2\vert X_1 = 3) &= \frac{1}{30}\frac{1}{35}[35(1) + 30(2) + 20(4) + 140(4) + 120(8) + 80(16)] \\
&= \frac{2{,}975}{1{,}050}
\end{aligned}
$$

(h)
$$\Pr(X_2 = 0, \ X_1 = 3) = \left[\frac{1}{6}\frac{35(1)}{900} + \frac{25(2)}{900} + \frac{25(4)}{900}\right.$$
$$\left. + \frac{10(4)}{900} + \frac{10(8)}{900} + \frac{10(16)}{900}\right] = \frac{455}{5,400}.$$

$\Pr(X_2 = 3 | X_1 = 3) = \frac{357}{5,400}$. $\Pr(X_2 = 8 | X_1 = 3) = \frac{238}{5,400}$.
(i) Divide answers to (h) by $\Pr(X_1 = 3) = 35/180$ to obtain the answers to (f).
(j) $E(X_2 | X_1 = 3) = 0\frac{455}{1050} + 3\frac{357}{1050} + 8\frac{238}{1050} = \frac{2,975}{1,050}$
(k)

$$\mu = \frac{1}{6}\frac{1}{30}(35 + 30 + 20 + 140 + 120 + 80) = \frac{425}{180}$$

$$v = \frac{1}{6}\frac{1}{900}(5{,}725 + 5{,}400 + 2{,}600 + 12{,}200 + 10{,}800 + 5{,}600) = \frac{43{,}325}{5{,}400}$$

$$a = \frac{1}{6}\frac{1}{900}(35^2 + 30^2 + 20^2 + 140^2 + 120^2 + 80^2) - \left(\frac{425}{180}\right)^2 = \frac{76{,}925}{32{,}400}$$

(l)

$$Z = \left(1 + \frac{43{,}425/5{,}400}{76{,}925/32{,}400}\right)^{-1}$$
$$= 0.228349.$$
$$P_c = 0.228349(3) + 0.771651(425/180)$$
$$= 2.507.$$

5.48 (a) $\pi(\theta_i) = 1/3$, $i = 1, 2, 3$.
(b)(c) The calculations appear in Table 4.4.

Table 4.4   Calculations for Exercise 5.48(b) and (c)

i	1	2	3
$\Pr(X = 0 | \theta_i)$	0.1600	0.0626	0.2500
$\Pr(X = 1 | \theta_i)$	0.2800	0.0500	0.1500
$\Pr(X = 2 | \theta_i)$	0.3225	0.3350	0.3725
$\Pr(X = 3 | \theta_i)$	0.1750	0.1300	0.1050
$\Pr(X = 4 | \theta_i)$	0.0625	0.4225	0.1225
$\mu(\theta_i)$	1.7	2.8	1.7
$v(\theta_i)$	1.255	1.480	1.655

(d) $\Pr(X_1 = 2) = \frac{1}{3}(0.3225 + 0.335 + 0.3725) = 0.34333$.
(e) The calculations appear in Table 4.5.
(f) The calculations appear in Table 4.6.

Table 4.5 Calculations for Exercise 5.48(e)

i	$\Pr(X_1 = 2\lvert\theta_i)$	$\Pr(\Theta = \theta_i\lvert X_1 = 2) = \frac{\Pr(X_1=2\lvert\theta_i)(1/3)}{.34333}$
1	0.3225	0.313107
2	0.3350	0.325243
3	0.3725	0.361650

Table 4.6 Calculations for Exercise 5.48(f)

x_2	$\Pr(X_2 = x_2\lvert X_1 = 2) = \sum \Pr(X_2 = x_2\lvert\Theta = \theta_i)\Pr(\Theta = \theta_i\lvert X_1 = 2)$
0	$0.16(0.313107) + 0.0625(0.325243) + 0.25(0.361650) = 0.160837$
1	$0.28(0.313107) + 0.05(0.325243) + 0.15(0.361650) = 0.158180$
2	$0.3225(0.313107) + 0.335(0.325243) + 0.3725(0.361650) = 0.344648$
3	$0.175(0.313107) + 0.13(0.325243) + 0.105(0.361650) = 0.135049$
4	$0.0625(0.313107) + 0.4225(0.325243) + 0.1225(0.361650) = 0.201286$

(g) $\begin{aligned} E(X_2\lvert X_1 = 2) &= 1.7(0.313107) + 2.8(0.325243) + 1.7(0.361650) \\ &= 2.057767. \end{aligned}$

(h) $\begin{aligned} \Pr(X_2 = 0, \; X_1 = 2) &= [0.16(0.3225) + 0.0625(0.335) + 0.25(0.3725)]/3 \\ &= 0.055221 \end{aligned}$

$\begin{aligned} \Pr(X_2 = 1\lvert X_1 = 2) &= 0.054308. \; \Pr(X_2 = 2\lvert X_1 = 2) = 0.344648 \\ \Pr(X_2 = 3\lvert X_1 = 2) &= 0.135049. \; \Pr(X_2 = 4\lvert X_1 = 2) = 0.069108. \end{aligned}$

(i) Divide answers to (h) by $\Pr(X_1 = 2) = 3.34333$ to obtain the answers to (f).

(j) $\begin{aligned} E(X_2\lvert X_1 = 2) &= 0(0.160837) + 1(0.158180) + 2(0.344648) + 3(0.135049) \\ &\quad +4(0.201286) = 2.057767. \end{aligned}$

(k) $\qquad \mu = \frac{1}{3}(1.7 + 2.8 + 1.7) = 2.06667$

$\qquad v = \frac{1}{3}(1.255 + 1.48 + 1.655) = 1.463333$

$\qquad a = \frac{1}{3}(1.7^2 + 2.8^2 + 1.7^2) - 2.06667^2 = 0.268889.$

(l) $\qquad \begin{aligned} Z &= \left(1 + \frac{1.463333}{.267779}\right)^{-1} \\ &= 0.155228. \\ P_c &= 0.155228(2) + 0.844772(2.06667) \\ &= 2.056321 \end{aligned}$

(m) Table 4.4 becomes Table 4.7 and the quantities become $\mu = 1.033333$, $v = 0.731667$, $\alpha = 0.067222$. $Z = \frac{2}{2+.731667/0.067222} = 0.155228$.

Table 4.7   Calculations for Exercise 5.48(m)

i	$\Pr(X=0|\theta_i)$	$\Pr(X=1|\theta_i)$	$\Pr(X=2|\theta_i)$	$\mu(\theta_i)$	$v(\theta_i)$
1	0.40	0.35	0.25	0.85	0.6275
2	0.25	0.10	0.65	1.40	0.7400
3	0.50	0.15	0.35	0.85	0.8275

5.49

$$E(S|\theta_A) = E(N|\theta_A)E(X|\theta_A) = 0.2(200) = 40$$
$$E(S|\theta_B) = 0.7(100) = 70$$
$$Var(S|\theta_A) = E(N|\theta_A)Var(X|\theta_A) + Var(N|\theta_A)[E(X|\theta_A)]^2$$
$$= 0.2(400) + 0.2(40,000) = 8,800$$
$$Var(S|\theta_B) = 0.7(1500) + 0.3(10,000) = 4,050$$
$$\mu_S = \frac{2}{3}40 + \frac{1}{3}70 = 50$$
$$v_S = \frac{2}{3}8,800 + \frac{1}{3}4,050 = 7,216.67$$
$$a_S = Var[\mu(\theta)] = \frac{2}{3}40^2 + \frac{1}{3}70^2 - 50^2 = 200$$
$$k = \frac{v_S}{a_S} = 36.083$$
$$Z = \frac{4}{4+36.083} = 0.10$$
$$P_c = 0.10(125) + 0.90(50) = 57.50$$

5.50 Let S denote total claims. Then $\mu_S = \mu_N \mu_Y = 0.1(100) = 10$.

$$v_S = E\{E(N|\theta_1)Var(Y|\theta_2) + Var(N|\theta_1)E[(Y|\theta_2)]^2\}$$
$$= E[E(N|\theta_1)]E[Var(Y|\theta_2)] + E[Var(N|\theta_1)]E\{[E(Y|\theta_2)]^2\}$$
$$= \mu_N v_Y + v_N E\{[E(Y|\theta_2)]^2\}$$

Since $a_Y = Var[\mu_Y(\theta)] = E\{[\mu_Y(\theta)]^2\} - \{E[\mu_Y(\theta)]\}^2$, $v_S = \mu_N v_Y + v_N(a_Y + \mu_Y^2)$. Then $a_Y = Var[E(Y|\theta_2)] = Var(\theta_2)$ since Y is exponentially distributed. But, again using the exponential distribution

$$Var(\theta_2) = E(\theta_2^2) - [E(\theta_2)]^2 = E[Var(Y|\theta_2)] - \{E[E(Y|\theta_2)]\}^2$$
$$= v_Y - \mu_Y^2,$$

from which $a_Y + \mu_Y^2 = v_Y$. Then

$$
\begin{aligned}
v_S &= \mu_N v_Y + v_N(v_Y - \mu_Y^2 + \mu_Y^2) = \mu_N v_Y + v_N v_Y \\
&= 0.1(25{,}000) + 0.1(25{,}000) = 5{,}000.
\end{aligned}
$$

Also,

$$
\begin{aligned}
a_S &= Var[\mu_S(\theta_1, \theta_2)] = E\{[\mu_S(\theta_1, \theta_2)]^2\} - \{E[\mu_S(\theta_1, \theta_2)]\}^2 \\
&= E\{[\mu_N(\theta_1)]^2\} E\{[(\mu_Y(\theta_2)]^2\} - \mu_N^2 \mu_Y^2 \\
&= (a_N + \mu_N^2)(a_Y + \mu_Y^2) - \mu_N^2 \mu_Y^2 \\
&= [0.05 + (0.01)^2](25{,}000) - (0.1)^2(100)^2 = 1{,}400.
\end{aligned}
$$

Therefore, $k = v_S/a_S = 5{,}000/1500 = 3.33$, $Z = \frac{3}{3+3.33} = 0.474$ and $P_c = 0.474\left(\frac{200}{3}\right) + 0.526(10) = 36.84$.

5.51 (a) $E(X_j) = E[E(X_j|\Theta)] = E[\beta_j \mu(\Theta)] = \beta_j E[\mu(\Theta)] = \beta_j \mu$.

$$
\begin{aligned}
Var(X_j) &= E[Var(X_j|\Theta)] + Var[E(X_j|\Theta)] \\
&= E[\tau_j(\Theta) + \psi_j v(\Theta)] + Var[\beta_j \mu(\Theta)] \\
&= \tau_j + \psi_j v + \beta_j^2 a.
\end{aligned}
$$

$$
\begin{aligned}
Cov(X_i, X_j) &= E(X_i X_j) - E(X_i)E(X_j) \\
&= E[E(X_i X_j|\Theta)] - \beta_i \beta_j \mu^2 \\
&= E[E(X_i|\Theta)\, E(X_j|\Theta)] - \beta_i \beta_j \mu^2 \\
&= E\{\beta_i \beta_j [\mu(\Theta)]^2\} - \beta_i \beta_j \mu^2 \\
&= \beta_i \beta_j (E\{[\mu(\Theta)]^2\} - \{E[\mu(\Theta)]\}^2) \\
&= \beta_i \beta_j a.
\end{aligned}
$$

(b) The normal equations are

$$
E(X_{n+1}) = \tilde{\alpha}_0 + \sum_{j=1}^{n} \tilde{\alpha}_j E(X_j)
$$

$$
Cov\,(X_i, X_{n+1}) = \sum_{j=1}^{n} \tilde{\alpha}_j Cov(X_i, X_j)
$$

where $E(X_{n+1}) = \beta_{n+1}\mu$, $E(X_j) = \beta_j \mu$ and $Cov\,(X_i, X_{n+1}) = \beta_i \beta_{n+1} a$, and $Cov\,(X_i, X_i) = Var\,(X_i) = \tau_j + \psi_j v + \beta_j^2 a$. On substitution

$$
\beta_{n+1}\mu = \tilde{\alpha}_0 + \sum_{j=1}^{n} \tilde{\alpha}_j \beta_j \mu \tag{4.1}
$$

and

$$\beta_i \beta_{n+1} a = \sum_{j=1}^{n} \tilde{\alpha}_j \beta_i \beta_j a + \tilde{\alpha}_1 (\tau_i + \psi_i v)$$

$$= \left(\beta_{n+1} - \frac{\tilde{\alpha}_0}{\mu} \right) \beta_i a + \tilde{\alpha}_i (\tau_i + \psi_i v) \text{ from (4.1).}$$

Hence,

$$\frac{\tilde{\alpha}_0}{\mu} \beta_i a = \tilde{\alpha}_i (\tau_i + \psi_i v),$$

yielding

$$\tilde{\alpha}_i = \frac{\tilde{\alpha}_0}{\mu} \beta_i a (\tau_i + \psi_i v)^{-1}. \tag{4.2}$$

Then

$$\sum_{i=1}^{n} \tilde{\alpha}_i = \frac{\tilde{\alpha}_0}{\mu} a \sum_{i=1}^{n} (\tau_i + \psi_i v)^{-1}.$$

From (4.1) and (4.2),

$$\tilde{\alpha}_0 = \beta_{n+1}\mu - \sum_{j=1}^{n} \tilde{\alpha}_j \beta_j \mu = \beta_{n+1}\mu - \tilde{\alpha}_0 a \sum_{j=1}^{n} \frac{\beta_j^2}{\tau_j + \psi_j v}$$

which gives

$$\tilde{\alpha}_0 = \frac{\beta_{n+1}\mu}{1 + a \sum_{j=1}^{n} \frac{\beta_j^2}{\tau_j + \psi_j v}} = \frac{\beta_{n+1}\mu}{1 + a \sum_{j=1}^{n} m_j} = \frac{\beta_{n+1}\mu}{1 + am}$$

and

$$\tilde{\alpha}_i = \frac{\beta_i \beta_{n+1}}{1 + am} \frac{a}{\tau_i + \psi_i v}.$$

The credibility premium is

$$\tilde{\alpha}_0 + \sum_{j=1}^{n} \tilde{\alpha}_j X_j = \frac{\beta_{n+1}\mu}{1 + am} + \frac{\beta_{n+1}a}{1 + am} \sum_{j=1}^{n} \frac{\beta_j}{\tau_j + \psi_j v} X_j$$

$$= \frac{E(X_{n+1})}{1 + am} + \frac{\beta_{n+1}a}{1 + am} \sum_{j=1}^{n} \frac{m_j}{\beta_j} X_j$$

$$= \frac{E(X_{n+1})}{1 + am} + \frac{\beta_{n+1}am}{1 + am} \overline{X}$$

$$= (1 - Z)E(X_{n+1}) + Z\beta_{n+1}\overline{X}.$$

5.52 The posterior distribution is

$$\pi(\theta|\mathbf{x}) \propto \left[\prod_{j=1}^{n} f(x_j|\theta)\right]\pi(\theta) \propto \prod_{j=1}^{n}\left[\theta^{x_j}(1-\theta)^{K_j-x_j}\right]\theta^{a-1}(1-\theta)^{b-1}$$

$$= \theta^{\Sigma x_j+a-1}(1-\theta)^{\Sigma(K_j-x_j)+b-1}$$

$$= \theta^{a_*}(1-\theta)^{b_*}$$

which is the kernel of the beta distribution with parameters $a_* = \sum x_j + a$ and $b_* = \sum(K_j - x_j) + b$. So

$$
\begin{aligned}
E(X_{n+1}|\mathbf{x}) &= \int_0^1 \mu_{n+1}(\theta)\pi(\theta|x)d\theta \\
&= \int_0^1 K_{n+1}\theta\frac{\Gamma(a_*+b_*)}{\Gamma(a_*)\Gamma(b_*)}\theta^{a_*-1}(1-\theta)^{b_*-1}d\theta \\
&= K_{n+1}\frac{\Gamma(a_*+b_*)}{\Gamma(a_*)\Gamma(b_*)}\int_0^1\theta^{a_*+1-1}(1-\theta)^{b_*-1}d\theta \\
&= K_{n+1}\frac{\Gamma(a_*+b_*)}{\Gamma(a_*)\Gamma(b_*)}\frac{\Gamma(a_*+1)\Gamma(b_*)}{\Gamma(a_*+1+b_*)} \\
&= K_{n+1}\frac{a_*}{a_*+b_*} \\
&= K_{n+1}\frac{\sum x_j+a}{\sum x_j+a+\sum(K_j-x_j)+b} \\
&= K_{n+1}\left(\frac{\sum x_j}{\sum K_j+a+b}\frac{\sum K_j}{\sum K_j}+\frac{a}{\sum K_j+a+b}\frac{a+b}{a+b}\right) \\
&= K_{n+1}\left(\frac{\sum x_j}{\sum K_j}\frac{\sum K_j}{\sum K_j+a+b}+\frac{a}{a+b}\frac{a+b}{\sum K_j+a+b}\right)
\end{aligned}
$$

Let

$$Z = \frac{\Sigma K_j}{\Sigma K_j+a+b}, \quad \overline{X} = \frac{\Sigma X_j}{\Sigma K_j}, \quad \mu = \frac{a}{a+b}.$$

Then $(EX_{n+1}|\mathbf{x}) = K_{n+1}[Z\overline{X} + (1-Z)\mu]$. Normalizing, $E\left(\frac{X_{n+1}}{K_{n+1}}\Big|\mathbf{x}\right) = Z\overline{X} + (1-Z)\mu$, the credibility premium.

5.53 The posterior distribution is

$$\pi(\theta|\mathbf{x}) \propto \left[\prod_{j=1}^{n} f(x_j|\theta)\right]\pi(\theta) \propto \theta^n e^{-\theta\Sigma x_j}\theta^{\alpha-1}e^{-\theta/\beta} = \theta^{n+\alpha-1}e^{-\theta\left(\Sigma x_j+\beta^{-1}\right)}$$

which is the kernel of the gamma distribution. So, noting that $\mu(\theta) = \theta^{-1}$,

$$E(X_{n+1}|\mathbf{x}) = \int_0^\infty \mu_{n+1}(\theta)\pi(\theta|\mathbf{x})d\theta$$

$$= \frac{\left(\sum x_j + \beta^{-1}\right)^{n+\alpha}}{\Gamma(n+\alpha)} \int_0^\infty \frac{1}{\theta} \theta^{n+\alpha-1} e^{-\theta\left(\Sigma x_j + \beta^{-1}\right)} d\theta$$

$$= \frac{\left(\sum x_j + \beta^{-1}\right)^{n+\alpha}}{\Gamma(n+\alpha)} \int_0^\infty \theta^{n+\alpha-2} e^{-\theta\left(\Sigma x_j + \beta^{-1}\right)} d\theta$$

$$= \frac{\left(\sum x_j + \beta^{-1}\right)^{n+\alpha}}{\Gamma(n+\alpha)} \frac{\Gamma(n+\alpha-1)}{\left(\sum x_j + \beta^{-1}\right)^{n+\alpha-1}}$$

$$= \frac{\sum x_j + \beta^{-1}}{n+\alpha-1}$$

$$= \frac{n\beta}{(n+\alpha-1)\beta}\bar{x} + \left(1 - \frac{n\beta}{(n+\alpha-1)\beta}\right)\frac{1}{\beta(\alpha-1)}$$

$$= Z\bar{x} + (1-Z)\mu$$

where

$$\mu = \int_0^\infty \frac{1}{\theta} \frac{\theta^{\alpha-1} e^{-\theta/\beta}}{\Gamma(\alpha)\beta^\alpha} d\theta = \frac{1}{\beta(\alpha-1)}.$$

5.54 The posterior distribution is

$$\pi(\theta|\mathbf{x}) \propto \left[\prod_{j=1}^n f(x_j|\theta)\right] \pi(\theta)$$

$$\propto \prod_{j=1}^n [\theta^r (1-\theta)^{x_j}] \theta^{a-1}(1-\theta)^{b-1}$$

$$= \theta^{nr+a-1}(1-\theta)^{\Sigma x_j + b - 1}$$

which is the kernel of the beta distribution.

$$E(X_{n+1}|\mathbf{x}) = \int_0^1 \mu(\theta)\pi(\theta|\mathbf{x})d\theta$$

$$= \int_0^1 r\frac{1-\theta}{\theta} \frac{\Gamma(nr + a + \sum x_j + b)}{\Gamma(nr+a)\Gamma(\sum x_j + b)} \theta^{nr+a-1}(1-\theta)^{\Sigma x_j + b - 1} d\theta$$

$$= r\frac{\Gamma(nr + a + \sum x_j + b)}{\Gamma(nr+a)\Gamma(\sum x_j + b)} \frac{\Gamma(nr + a - 1)\Gamma(\sum x_j + b + 1)}{\Gamma(nr + a + \sum x_j + b)}$$

$$= r\frac{\sum x_j + b}{nr + a - 1} = \frac{nr}{nr + a - 1}\bar{x} + \left(1 - \frac{nr}{nr + a - 1}\right)\frac{rb}{a-1}.$$

But,

$$\mu = E[E(X|\theta)] = \int_0^1 r\frac{1-\theta}{\theta} \frac{\Gamma(a+b)}{\Gamma(a)\Gamma(b)} \theta^{a-1}(1-\theta)^{b-1} d\theta = \frac{rb}{a-1}$$

and so $(EX_{n+1}|\mathbf{x}) = Z\overline{X} + (1-Z)\mu$ where $Z = \frac{nr}{nr+a-1}$.

5.55 (a) The posterior distribution is

$$
\begin{aligned}
\pi(\theta|\mathbf{x}) &\propto \left[\prod_{j=1}^{n} \frac{\exp(-m\theta x_j)}{[q(\theta)]^m} \right] [q(\theta)]^{-k} \exp(-\theta\mu k) \\
&= [q(\theta)]^{-k-mn} \exp\left[-\theta \left(m\sum x_j + \mu k \right) \right]
\end{aligned}
$$

This is of the same form as the prior distribution, with

$$
k_* = k + mn \text{ and } \mu_* = \frac{m\sum x_j + \mu k}{mn + k}
$$

The Bayesian premium is

$$
\begin{aligned}
E(X_{n+1}|\mathbf{x}) &= -\int_{\theta_0}^{\theta_1} \frac{q'(\theta)}{q(\theta)} [q(\theta)]^{-k_*} \exp(-\theta\mu_* k_*) c(\mu_*, k_*)^{-1} d\theta \\
&= -\int_{\theta_0}^{\theta_1} q'(\theta)[q(\theta)]^{-k_*-1} \exp(-\theta\mu_* k_*) c(\mu_*, k_*)^{-1} d\theta \\
&= -[q(\theta)]^{-k_*} \exp(-\theta\mu_* k_*) c(\mu_*, k_*)^{-1} \Big|_{\theta_0}^{\theta_1} \\
&\quad + \int_{\theta_0}^{\theta_1} q(\theta)\{-(k_*+1)[q(\theta)]^{-k_*-2}q'(\theta) \exp(-\theta\mu_* k_*) \\
&\quad -[q(\theta)]^{-k_*-1} \exp(-\theta\mu_* k_*)\mu_* k_*\} c(\mu_*, k_*)^{-1} d\theta \\
&= 0 + (k_*+1)E(X_{n+1}|\mathbf{x}) + \mu_* k_*.
\end{aligned}
$$

and so

$$
E(X_{n+1}|\mathbf{x}) = \mu_* = \frac{m\sum x_j + \mu k}{mn + k} = \frac{mn}{mn + k}\bar{x} + \frac{k}{mn + k}\mu.
$$

(b) The Bühlmann premium must be the same because the Bayesian premium is linear in the observations.

(c) The inverse Gaussian distribution (see Appendix A) can be written

$$
f(x) = \left(\frac{\theta}{2\pi x^3} \right)^{1/2} \exp\left(-\frac{\theta x}{2\mu^2} + \frac{\theta}{\mu} - \frac{\theta}{2x} \right).
$$

Replace θ with m and μ with $(2\theta)^{-1/2}$ to obtain

$$
f(x) = \left(\frac{m}{2\pi x^3} \right)^{1/2} \exp\left[-m\theta x + m(2\theta)^{1/2} - \frac{m}{2x} \right].
$$

Now let $p(m, x) = \left(\frac{m}{2\pi x^3} \right)^{1/2} \exp\left(\frac{m}{2x} \right)$ and $q(\theta) = \exp[-(2\theta)^{1/2}]$ to see that $f(x)$ has the desired form.

5.56 (a) It is true when $X_1, X_2, \ldots$ represent values in successive years and τ is an inflation factor for each year.

(b) This is a special case of Exercise 5.51 with $\beta_j = \tau^j$, $\tau_j(\theta) = 0$, $\psi_j = \tau^{2j}/m_j$ for $j = 1, \ldots, n$.

(c) From Exercise 5.51(b)

$$\tilde{\alpha}_0 + \sum_{j=1}^{n} \tilde{\alpha}_j X_j = (1 - Z)E(X_{n+1}) + Z\tau^{n+1}\overline{X}$$

where

$$\overline{X} = \sum_{j=1}^{n} \frac{\tau^{2j}}{\frac{\tau^{2j}}{m_j}} \frac{X_j}{\tau_j} \bigg/ \left[\sum_{j=1}^{m} \frac{\tau^{2j}}{\frac{\tau^{2j}}{m_j}}\right] = \sum_{j=1}^{n} \frac{m_j X_j}{\tau^j} \bigg/ \sum_{j=1}^{n} m_j = \sum_{j=1}^{n} \frac{m_j}{m} \frac{X_j}{\tau_j}$$

and

$$Z = \frac{a \sum_{j=1}^{n} \frac{m_j}{v}}{1 + a \sum_{j=1}^{n} \frac{m_j}{v}} = \frac{m}{k+m}.$$

Then

$$\tilde{\alpha}_0 + \sum_{j=1}^{n} \tilde{\alpha}_j X_j = \frac{k}{k+m}E(X_{n+1}) + \frac{m}{k+m}\tau^{n+1}\overline{X}$$

$$= \frac{k}{k+m}\tau^{n+1}\mu + \frac{m}{k+m}\sum_{j=1}^{n} \frac{m_j}{m}\tau^{n-1-j}X_j.$$

(d) As in Exercise 5.51, the credibility premium equals the Bayesian premium.

(e)
$$\frac{\frac{\partial}{\partial\theta}f(x_j|\theta)}{f(x_j|\theta)} = -m_j\tau^{-j}x_j - m_j\frac{q'(\theta)}{q(\theta)}$$

Therefore,

$$\int \frac{\partial}{\partial\theta}f(x_j|\theta)dx_j = -m_j\tau^{-j}\int x_j f(x_j|\theta)dx_j + m_j\mu(\theta)\int f(x_j|\theta)dx_j$$

which leads to

$$0 = -m_j\tau^{-j}E(X_j|\theta) + m_j\mu(\theta)$$

so that $E(X_j|\theta) = \tau^j\mu(\theta)$. Also,

$$\frac{\partial^2}{\partial\theta^2}f(x_j|\theta) = m_j\mu'(\theta)f(x_j|\theta) + [m_j\tau^{-j}x_j - m_j\mu(\theta)]^2 f(x_j|\theta)$$

$$= m_j\mu'(\theta)f(x_j|\theta) + (m_j\tau^{-j})^2[x_j - \tau^j\mu(\theta)]^2 f(x_j|\theta).$$

Integrating leads to

$$0 = m_j\mu'(\theta) + (m_j\tau^{-j})^2 Var(X_j|\theta)$$

from which

$$Var(X_j|\theta) = \frac{\tau^{2j}}{m_j}v(\theta).$$

(f) As in Exercise 5.51, $E[\mu(\theta)] = \mu$. Then

$$\pi(\theta|\mathbf{x}) \quad \propto \quad \left[\prod_{j=1}^{n} f(x_j|\theta)\right]\pi(\theta) \propto \prod_{j=1}^{n}\left\{\frac{e^{-m_j\tau^{-j}x_j\theta}}{[q(\theta)]^{m_j}}\right\}[q(\theta)]^{-k}e^{-\theta\mu k}$$

$$= \quad [q(\theta)]^{-k-m}\exp\left[-\theta\left(\sum_{j=1}^{n}\tau^{-j}m_jx_j + \mu k\right)\right] = [q(x)]^{-k_*}e^{-\theta\mu_* k_*}$$

which is of the same form as the prior with

$$k_* = k + m \text{ and } k_*\mu_* = k\mu + \sum_{j=1}^{n}m_j\tau^{-j}X_j$$

so that

$$\mu_* = \frac{k}{k+m}\mu + \frac{m}{k+m}\sum_{j=1}^{n}\frac{m_j}{m}\tau^{-j}X_j.$$

Therefore $E[X_{n+1}|\mathbf{x}]$ is linear in the X_j's. Hence, credibility is exact.

5.57 (a) The Poisson *pgf* of each X_j is $P_{X_j}(z|\theta) = e^{-\theta(z-1)}$ and so the *pgf* of S is $P_S(z|\theta) = e^{-n\theta(z-1)}$ which is Poisson with mean $n\theta$. Hence,

$$f_{S|\theta}(s|\theta) = \frac{(n\theta)^s e^{-n\theta}}{s!}$$

and so

$$f_S(s) = \int \frac{(n\theta)^s e^{-n\theta}}{s!}\pi(\theta)d\theta.$$

(b) $\mu(\theta) = E(X|\theta) = \theta$ and $\pi(\theta|\mathbf{x}) = \left[\prod_{j=1}^{n} f(x_j|\theta)\right]\pi(\theta)/f(x)$. We have,

$$f(x) \quad = \quad \int\left[\prod_{j=1}^{n} f(x_j|\theta)\right]\pi(\theta)d\theta$$

$$= \quad \frac{\int\theta^{\Sigma x_j}e^{-n\theta}\pi(\theta)d\theta}{\prod_{j=1}^{n}x_j!}$$

Therefore,

$$\pi(\theta|\mathbf{x}) = \frac{\theta^{\Sigma x_j}e^{-n\theta}\pi(\theta)}{\int_0^{\infty}\theta^{\Sigma x_j}e^{-n\theta}\pi(\theta)d\theta}$$

The Bayesian premium is

$$E(X_{n+1}|\mathbf{x}) = \int_0^\infty \mu(\theta)\pi(\theta|x)d\theta = \frac{\int_0^\infty \theta^{\Sigma x_j+1}e^{-n\theta}\pi(\theta)d\theta}{\int_0^\infty \theta^{\Sigma x_j}e^{-n\theta}\pi(\theta)d\theta}$$

$$= \frac{\frac{(s+1)!}{n^{s+1}}\int_0^\infty \frac{n^{s+1}}{(s+1)!}\theta^{s+1}e^{-n\theta}\pi(\theta)d\theta}{\frac{s!}{n^s}\int_0^\infty \frac{n^s}{s!}\theta^s e^{-n\theta}\pi(\theta)d\theta} = \frac{s+1}{n}\frac{f_S(s+1)}{f_S(s)}.$$

(c)

$$f_S(s) = \int_0^\infty \frac{(n\theta)^s e^{-n\theta}}{s!}\frac{\beta^{-\alpha}}{\Gamma(\alpha)}\theta^{\alpha-1}e^{-\theta/\beta}d\theta$$

$$= \frac{n^s}{s!}\frac{\beta^{-\alpha}}{\Gamma(\alpha)}\int_0^\infty \theta^{s+\alpha-1}e^{-(n+\beta^{-1})\theta}d\theta$$

$$= \frac{n^s}{s!}\frac{\beta^{-\alpha}}{\Gamma(\alpha)}\frac{\Gamma(s+\alpha)}{(n+\beta^{-1})^{s+\alpha}}$$

$$= \frac{\Gamma(s+\alpha)}{\Gamma(s+1)\Gamma(\alpha)}\left(\frac{1}{1+n\beta}\right)^\alpha\left(\frac{n\beta}{1+n\beta}\right)^s$$

$$= \binom{r+s-1}{s}\left(\frac{1}{1+\beta_*}\right)^r\left(\frac{\beta_*}{1+\beta_*}\right)^s$$

where $\beta_* = n\beta$ and $r = \alpha$. This is a negative binomial distribution.

5.58 (a) Let Θ represent the selected urn. Then, $f_X(x|\Theta = 1) = \frac{1}{4}$, $x = 1, 2, 3, 4$ and $f_X(x|\Theta = 2) = \frac{1}{6}$, $x = 1, 2, \ldots, 6$. Then, $\mu(1) = E(X|\Theta = 1) = 2.5$ and $\mu(2) = E(X|\theta = 2) = 3.5$

For the Bayesian solution, the marginal probability of drawing a 4 is, $f_X(4) = \frac{1}{2}\times\frac{1}{4} + \frac{1}{2}\times\frac{1}{6} = \frac{5}{24}$ and the posterior probability for urn 1 is

$$\pi(\theta = 1|X = 4) = \frac{f_{X|\Theta}(4|1)\pi(1)}{f_X(4)} = \frac{\frac{1}{4}\frac{1}{2}}{\frac{5}{24}} = \frac{3}{5}$$

and for urn 2 is

$$\pi(\theta = 2|X = 4) = 1 - \frac{3}{5} = \frac{2}{5}.$$

The expected value of the next observation is

$$E(X_2|X_1 = 4) = 2.5\left(\frac{3}{5}\right) + 3.5\left(\frac{2}{5}\right) = 2.9.$$

(b) Using Bühlmann credibility,

$$\mu = \frac{1}{2}(2.5 + 3.5) = 3$$

$$v(1) = \frac{1}{4}(1^2 + 2^2 + 3^2 + 4^2) - (2.5)^2 = 1.25$$

$$v(2) = \frac{1}{6}(1^2 + \cdots + 6^2)(3.5)^2 = 2.917$$

$$v = \frac{1}{2}v(1) + \frac{1}{2}v(2) = 2.0835$$

$$a = Var[\mu(\theta)] = E\{[\mu(\theta)]^2\} - \{E[\mu(\theta)]\}^2 = \frac{1}{2}[2.5^2 + 3.5^2] - 3^2$$

$$= 0.25$$

$$k = \frac{v}{a} = \frac{2.0835}{0.25} = 8.334$$

$$Z = \frac{1}{1 + 8.334} = 0.1071352.$$

The credibility premium is $P_c = Z\bar{x} + (1-Z)\mu = 0.1071352(4) + 0.8927648(3) = 3.10709$.

5.59 (a)

$$\mu(\theta) = \theta, v(\theta) = \theta$$

$$\mu = E(\theta) = \int_1^\infty 3\theta^{-3}d\theta = 1.5$$

$$v = E(\theta) = 1.5$$

$$a = Var(\theta) = \int_1^\infty 3\theta^{-2}d\theta - 2.25 = 0.75$$

$$k = 1.5/0.75 = 2$$

$$Z = \frac{2}{2+2} = 0.5$$

$$P_c = 0.5(10) + 0.5(1.5) = 5.75.$$

(b) $\pi(\theta|N_1 + N_2 = 20) \propto e^{-2\theta}\theta^{20}\theta^{-4} = e^{-2\theta}\theta^{16}$. This is a gamma distribution with parameters 17 and 0.5. The mean is 8.5.

5.60 $Z = \frac{0.5}{0.5+k} = 0.5$, $k = 0.5$, $Z = \frac{3}{3+0.5} = 6/7$.

5.61 (a) $\Pr(X_1 = 1|A) = 3(0.1^2)(0.9) = 0.027$

$\Pr(X_1 = 1|B) = 3(0.6^2)(0.4) = 0.432$

$\Pr(X_1 = 1|C) = 3(0.8^2)(0.2) = 0.384$

$\Pr(A|X_1 = 1) = 0.027/(0.027 + 0.432 + 0.384) = 27/843 = 9/281$

$\Pr(B|X_1 = 1) = 144/281$

$$\begin{aligned}
\Pr(C|X_1 = 1) &= 128/281 \\
\mu(A) &= 3(0.9) = 2.7 \\
\mu(B) &= 3(0.4) = 1.2 \\
\mu(C) &= 3(0.2) = 0.6 \\
E(X_2|X_1 = 1) &= [9(2.7) + 144(1.2) + 128(0.6)]/281 = 0.97473.
\end{aligned}$$

(b)
$$\begin{aligned}
\mu &= (2.7 + 1.2 + 0.6)/3 = 1.5 \\
a &= (2.7^2 + 1.2^2 + 0.6^2)/3 - 2.25 = 0.78 \\
v(A) &= 3(0.9)(0.1) = 0.27, \; v(B) = 0.72, \; v(C) = 0.48 \\
v &= (0.27 + 0.72 + 0.48)/3 = 0.49 \\
k &= 49/78, \; Z = (1 + 49/78)^{-1} = 78/127 \\
P_c &= \frac{78}{127}(1) + \frac{49}{127}(1.5) = 1.19291.
\end{aligned}$$

5.62 (a)
$$\begin{aligned}
\mu(\lambda) &= \lambda, \; v(\lambda) = \lambda \\
\mu &= E(\lambda) = \int_1^\infty 4\lambda^{-4} d\lambda = 4/3 \\
v &= E(\lambda) = 4/3 \\
a &= Var(\lambda) = \int_1^\infty 4\lambda^{-3} d\lambda - 16/9 = 2/9 \\
k &= (4/3)/(2/9) = 6, \; Z = \frac{3}{3+6} = 1/3 \\
P_c &= (1/3)(1) + (2/3)(4/3) = 11/9
\end{aligned}$$

(b)
$$\begin{aligned}
\mu &= \int_0^1 \lambda d\lambda = 1/2 \\
v &= \mu = 1/2 \\
a &= \int_0^1 \lambda^2 d\lambda - 1/4 = 1/12 \\
k &= (1/2)/(1/12) = 6, \; Z = \frac{3}{3+6} = 1/3 \\
P_c &= (1/3)(1) + (2/3)(1/2) = 2/3.
\end{aligned}$$

5.63 $\mu(h) = h$, $\mu = E(h) = 2$, $v(h) = h$, $v = E(h) = 2$, $a = Var(h) = 2$, $k = 2/2 = 1$, $Z = \frac{1}{1+1} = 1/2$.

5.64 (a) $r \sim bin(3, \theta)$, $\pi(\theta) = 6\theta(1 - \theta)$.
$$\pi(\theta|X = 1) \propto 3\theta(1 - \theta)^2 6\theta(1 - \theta) \propto \theta^2(1 - \theta)^3$$

and so the posterior distribution is beta with parameters 3 and 4. Then the expected next observation is $E(3\theta|X=1) = 3(3/7) = 9/7$.

(b)
$$\mu(\theta) = 3\theta, \; v(\theta) = 3\theta(1-\theta)$$

$$\mu = E(3\theta) = 3\int_0^1 \theta 6\theta(1-\theta)d\theta = 1.5$$

$$v = E[3\theta(1-\theta)] = 3\int_0^1 \theta(1-\theta)6\theta(1-\theta)d\theta = 0.6$$

$$a = Var(3\theta) = 9\int_0^1 \theta^2 6\theta(1-\theta)d\theta - 2.25 = 0.45$$

$$k = 0.6/0.45 = 4/3, \; Z = (1+4/3)^{-1} = 3/7$$

$$P_c = (3/7)(1) + (4/7)(1.5) = 9/7.$$

5.65 (a)
$$\mu(A) = 20, \; \mu(B) = 12, \; \mu(C) = 10$$

$$v(A) = 416, \; v(B) = 288, \; v(C) = 308$$

$$\mu = (20+12+10)/3 = 14$$

$$v = (416+288+308)/3 = 337\frac{1}{3}$$

$$a = (20^2+12^2+10^2)/3 - 14^2 = 18\frac{2}{3}$$

$$k = 337\frac{1}{3}/18\frac{2}{3} = 18\frac{1}{14}$$

$$Z = (1+18\frac{1}{14})^{-1} = 14/267$$

$$P_c = (14/267)(0) + (253/267)(14) = 13.2659.$$

(b)
$$\pi(A|X=0) = 2/(2+3+4) = 2/9$$

$$\pi(B|X=0) = 3/9, \; \pi(C|X=0) = 4/9$$

$$E(X_2|X_1=0) = (2/9)20 + (3/9)12 + (4/9)10 = 12\frac{8}{9}.$$

5.66 (a) $\Pr(N=0) = \int_1^3 e^{-\lambda}(0.5)d\lambda = (e^{-1} - e^{-3})/2 = 0.159046$.

(b)
$$\mu = v = E(\lambda) = \int_1^3 \lambda(0.5)d\lambda = 2$$

$$a = Var(\lambda) = \int_1^3 \lambda^2(0.5)d\lambda - 4 = 1/3$$

$$k = 2/(1/3) = 6, \; Z = \frac{1}{1+6} = 1/7$$

$$P_c = (1/7)(1) + (6/7)(2) = 13/7.$$

(c) $\pi(\lambda|X_1 = 1) = e^{-\lambda}\lambda(.5)/\int_1^3 e^{-\lambda}\lambda(.5)d\lambda = e^{-\lambda}\lambda/(2e^{-1} - 4e^{-3})$,

$$\begin{aligned} E(\lambda|X_1 = 1) &= \int_1^3 e^{-\lambda}\lambda^2 d\lambda/(2e^{-1} - 4e^{-3}) \\ &= (5e^{-1} - 17e^{-3})/(2e^{-1} - 4e^{-3}) = 1.8505. \end{aligned}$$

5.67 (a)
$$\begin{aligned} \mu(A) &= (1/6)(4) = 2/3, \ \mu(B) = (5/6)(2) = 4/3 \\ v(A) &= (1/6)(20) + (5/36)(16) = 50/9 \\ v(B) &= (5/6)(5) + (5/36)(4) = 85/18 \\ \mu &= [(2/3) + (4/3)]/2 = 1 \\ v &= [(50/9) + (85/18)]/2 = 185/36 \\ a &= [(2/3)^2 + (4/3)^2]/2 - 1 = 1/9 \\ k &= (185/36)/(1/9) = 185/4 \\ Z &= \frac{4}{4 + 185/4} = 16/201. \end{aligned}$$

(b) $(16/201)(.25) + (185/201)(1) = 189/201 = 0.9403.$

5.68
$$\begin{aligned} E(X_2) &= (1 + 8 + 12)/3 = 7 \\ &= E[E(X_2|X_1)] \\ &= [2.6 + 7.8 + E(X_2|X_1 = T)]/3. \\ E(X_2|X_1 = T) &= 10.6. \end{aligned}$$

5.69 (a) $X \sim \text{Poisson}(\lambda)$, $\pi(\lambda) = e^{-\lambda/2}/2$. The posterior distribution with three claims is proportional to $e^{-\lambda}\lambda^3 e^{-\lambda/2} = \lambda^3 e^{-1.5\lambda}$ which is gamma with parameters 4 and $1/1.5$. The mean is $4/1.5 = 2\frac{2}{3}$.
(b)
$$\begin{aligned} \mu(\lambda) &= v(\lambda) = \lambda \\ \mu &= v = E(\lambda) = 2 \\ a &= Var(\lambda) = 4 \\ k &= 2/4 = 0.5, \ Z = \frac{1}{1 + 0.5} = \frac{2}{3} \\ P_c &= \frac{2}{3}(3) + \frac{1}{3}(2) = \frac{8}{3} = 2\frac{2}{3}. \end{aligned}$$

5.70 (a) $r \sim bin(3, \theta)$, $\pi(\theta) = 280\theta^3(1 - \theta)^4$ which is beta(4,5).

$$\pi(\theta|X = 2) \propto 3\theta^2(1 - \theta)280\theta^3(1 - \theta)^4 \propto \theta^5(1 - \theta)^5$$

and so the posterior distribution is beta with parameters 6 and 6. Then the expected next observation is $E(3\theta|X = 2) = 3(6/12) = 1.5$.

(b) $\quad \mu(\theta) = 3\theta, \; v(\theta) = 3\theta(1 - \theta)$

$$\mu = E(3\theta) = 3(4/9) = 4/3$$

$$v = E[3\theta(1 - \theta)] = 3 \int_0^1 \theta(1 - \theta)280\theta^3(1 - \theta)^4 d\theta$$

$$= 840 \frac{\Gamma(5)\Gamma(6)}{\Gamma(11)} = 2/3$$

$$a = Var(3\theta) = 9 \int_0^1 \theta^2 280\theta^3(1 - \theta)^4 d\theta - 16/9$$

$$= 2{,}520 \frac{\Gamma(6)\Gamma(5)}{\Gamma(11)} - 16/9 = 2/9$$

$$k = (2/3)/(2/9) = 3$$
$$Z = (1 + 3)^{-1} = 1/4$$
$$P_c = (1/4)(2) + (3/4)(4/3) = 1.5.$$

5.71 (a) $\quad \mu(A_1) = 0.15, \; \mu(A_2) = 0.05$

$$v(A_1) = 0.1275, \; v(A_2) = 0.0475$$

$$\mu = (0.15 + 0.05)/2 = 0.1$$

$$v = (0.1275 + 0.0475)/2 = 0.0875$$

$$a = (0.15^2 + 0.05^2)/2 - 0.1^2 = 0.0025$$

$$k = 0.0875/0.0025 = 35$$

$$Z = \frac{3}{3 + 35} = 3/38,$$

estimated frequency is $(3/38)(1/3) + (35/38)(0.1) = 9/76$.

$$\mu(B_1) = 24, \; \mu(B_2) = 34$$

$$v(B_1) = 64, \; v(B_2) = 84$$

$$\mu = (24 + 34)/2 = 29$$

$$v = (64 + 84)/2 = 74$$

$$a = (24^2 + 34^2)/2 - 29^2 = 25$$

$$k = 74/25$$

$$Z = \frac{1}{1 + 74/25} = 25/99,$$

estimated severity is $(25/99)(20) + (74/99)(29) = 294/11$.

The estimated total is $(9/76)(294/11) = 1323/418 = 3.1651$.

(b) Information about the various spinner combinations is given in Table 4.8

Table 4.8 Calculations for Exercise 5.71

Spinners	μ	v
A_1, B_1	3.6	83.04
A_1, B_2	5.1	159.99
A_2, B_1	1.2	30.56
A_2, B_2	1.7	59.11

$$\begin{aligned}
\mu &= (3.6 + 5.1 + 1.2 + 1.7)/4 = 2.9 \\
v &= (83.04 + 159.99 + 30.56 + 59.11)/4 = 83.175 \\
a &= (3.6^2 + 5.1^2 + 1.2^2 + 1.7^2)/4 - 2.9^2 = 2.415 \\
k &= 83.175/2.415 = 34.441, \ Z = \frac{3}{3 + 34.441} = 0.080126,
\end{aligned}$$

estimated total is $(0.080126)(20/3) + 0.919874(2.9) = 3.2018$.
(c) For part (a),

$$\begin{aligned}
\Pr(1|A_1) &= 3(0.15)(0.85)^2 = 0.325125 \\
\Pr(1|A_2) &= 3(0.05)(0.95)^2 = 0.135375 \\
\Pr(A_1|1) &= \frac{0.325125}{0.325125 + 0.135375} = 0.706026 \\
\Pr(A_2|1) &= 1 - 0.706026 = 0.293974.
\end{aligned}$$

Estimated frequency is $0.706026(0.15) + 0.293974(0.05) = 0.120603$.

$$\begin{aligned}
\Pr(20|B_1) &= 0.8 \\
\Pr(20|B_2) &= 0.3 \\
\Pr(B_1|20) &= \frac{0.8}{0.8 + 0.3} = 8/11 \\
\Pr(B_2|20) &= 3/11.
\end{aligned}$$

Estimated severity is $(8/11)(24) + (3/11)(34) = 26.727272$.
   Estimated total is $0.120603(26.727272) = 3.2234$.
   For part (b),

$$\begin{aligned}
\Pr(0, 20, 0|A_1, B_1) &= (0.85)^2(0.12) = 0.0867 \\
\Pr(0, 20, 0|A_1, B_2) &= (0.85)^2(0.045) = 0.0325125 \\
\Pr(0, 20, 0|A_2, B_1) &= (0.95)^2(0.04) = 0.0361 \\
\Pr(0, 20, 0|A_2, B_2) &= (0.95)^2(0.015) = 0.0135375.
\end{aligned}$$

The posterior probabilities are 0.51347, 0.19255, 0.21380, 0.08017 and the estimated total is

$$0.51347(3.6) + 0.19255(5.1) + 0.21380(1.2) + 0.08017(1.7) = 3.2234.$$

(d)
$$\begin{aligned}
\Pr(X_1 = 0, \ldots, X_{n-1} = 0|A_1, B_1) &= (0.85)^{n-1} \\
\Pr(X_1 = 0, \ldots, X_{n-1} = 0|A_1, B_2) &= (0.85)^{n-1} \\
\Pr(X_1 = 0, \ldots, X_{n-1} = 0|A_2, B_1) &= (0.95)^{n-1} \\
\Pr(X_1 = 0, \ldots, X_{n-1} = 0|A_2, B_2) &= (0.95)^{n-1}
\end{aligned}$$

$$
\begin{aligned}
&E(X_n|X_1 = 0, \ldots, X_{n-1} = 0) \\
&= \frac{(0.85)^{n-1}(3.6) + (0.85)^{n-1}(5.1) + (0.95)^{n-1}(1.2) + (0.95)^{n-1}(1.7)}{(0.85)^{n-1} + (0.85)^{n-1} + (0.95)^{n-1} + (0.95)^{n-1}} \\
&= \frac{2.9 + 8.7(0.85/0.95)^{n-1}}{2 + 2(0.85/0.95)^{n-1}}
\end{aligned}
$$

and the limit as $n \to \infty$ is $2.9/2 = 1.45$.

5.72 $\Pr(X = 0.12|A) = \dfrac{1}{\sqrt{2\pi}(0.03)} \exp\left[-\dfrac{(0.12 - 0.1)^2}{2(0.0009)}\right] = 10.6483,$

$\Pr(X = 0.12|B) = (X = 0.12|C) = 0$ (actually, just very close to zero), so $\Pr(A|X = 0.12) = 1$. The Bayesian estimate is $\mu(A) = 0.1$.

5.73 $E(X|X_1 = 4) = 2 = Z(4) + (1 - Z)(1)$, $Z = 1/3 = \frac{1}{1+k}$, $k = 2 = v/a = 3/a$, $a = 1.5$.

5.74 $v = E(v) = 8$, $a = Var(\mu) = 4$, $k = 8/4 = 2$, $Z = \frac{3}{3+2} = 0.6$.

5.75 (a)

$$f(y) = \int_0^\infty \lambda^{-1} e^{-y/\lambda} 400\lambda^{-3} e^{-20/\lambda} d\lambda = 400 \int_0^\infty \lambda^{-4} e^{-(20+y)/\lambda} d\lambda.$$

Let $\theta = (20 + y)/\lambda$, $\lambda = (20 + y)/\theta$, $d\lambda = -(20 + y)/\theta^2 d\theta$, and so

$$f(y) = 400 \int_0^\infty (20 + y)^{-3} \theta^2 e^{-\theta} d\theta = 800(20 + y)^{-3}$$

which is Pareto with parameters 2 and 20 and so the mean is $20/(2-1) = 20$. (b) $\mu(\lambda) = \lambda$, $v(\lambda) = \lambda^2$. The distribution of λ is inverse gamma with $\alpha = 2$ and $\theta = 20$. Then $\mu = E(\lambda) = 20/(2-1) = 20$ and $v = E(\lambda^2)$ which does not exist. The Bühlmann estimate does not exist.

(c) $\pi(\lambda|15,25) \propto \lambda^{-1}e^{-15/\lambda}\lambda^{-1}e^{-25/\lambda}400\lambda^{-3}e^{-20/\lambda} \propto \lambda^{-5}e^{-60/\lambda}$ which is inverse gamma with $\alpha = 4$ and $\theta = 60$. The posterior mean is $60/(4-1) = 20$.

5.76

$$\mu(\theta) = \theta, \ v(\theta) = \theta(1-\theta)$$
$$a = Var(\theta) = 0.07$$
$$v = E(\theta - \theta^2) = E(\theta) - Var(\theta) - [E(\theta)]^2$$
$$= 0.25 - 0.07 - (0.25)^2 = 0.1175$$
$$k = 0.1175/0.07 = 1.67857,$$
$$Z = \frac{1}{1+1.67857} = 0.37333.$$

5.77 (a) means are 0, 2, 4, and 6 while the variances are all 9. Thus

$$\mu = (0+2+4+6)/4 =$$
$$v = 9, \ a = (0+4+16+36)/4 - 9 = 5$$
$$Z = \frac{1}{1+9/5} = 5/14 = 0.35714.$$

(b)(i) $v = 9$, $a = 20$, $Z = \frac{1}{1+9/20} = 20/29 = 0.68966$.
(b)(ii) $v = 3.24$, $a = 5$, $Z = \frac{1}{1+3.24/5} = 5/8.24 = 0.60680$.
(b)(iii) $v = 9$, $a = (4+4+100+100)/4 - 36 = 16$, $Z = \frac{1}{1+9/16} = 16/25 = 0.64$.
(b)(iv) $Z = \frac{3}{3+9/5} = 15/24 = 0.625$.
(b)(v) $a = 5$, $v = (9+9+2.25+2.25)/4 = 5.625$, $Z = \frac{2}{2+5.625/5} = 10/15.625 = 0.64$.
The answer is (i).

5.78 (a) Preliminary calculations are given in Table 4.9.

Table 4.9   Calculations for Exercise 5.78

Risk	100	1,000	20,000	μ	v
1	0.5	0.3	0.2	4,350	61,382,500
2	0.7	0.2	0.1	2,270	35,054,100

$$Pr(100|1) = 0.5, \ Pr(100|2) = 0.7$$
$$Pr(1|100) = \frac{0.5(2/3)}{0.5(2/3) + 0.7(1/3)} = 10/17, \ Pr(2|100) = 7/17.$$

Expected value is $(10/17)(4350) + (7/17)(2,270) = 3,493.53$.

(b) μ = $(2/3)(4{,}350) + (1/3)(2{,}270) = 3{,}656.33$
       v = $(2/3)(61{,}382{,}500) + (1/3)(35{,}054{,}100) = 52{,}606{,}366.67$
       a = $(2/3)(4{,}350)^2 + (1/3)(2{,}270)^2 - 3{,}656.33^2 = 963{,}859.91$
       k = $54.579,\ Z = 1/55.579 = 0.017992.$

Estimate is $0.017992(100) + 0.982008(3{,}656.33) = 3{,}592.34.$

5.79 (a) $v(\mu, \lambda)$ = $\mu(2\lambda^2) = 2\mu\lambda^2$
              v = $2E(\mu\lambda^2) = 2E(\mu)[Var(\lambda) + E(\lambda)^2]$
                 = $2(0.1)(640{,}000 + 1{,}000^2)$
                 = $328{,}000.$

(b) $\mu(\mu, \lambda)$ = $\mu\lambda$
         a = $Var(\mu\lambda) = E(\mu^2\lambda^2) - E(\mu)^2 E(\lambda)^2$
            = $[Var(\mu) + E(\mu)^2][Var(\lambda) + E(\lambda)^2] - E(\mu)^2 E(\lambda)^2$
            = $(0.0025 + 0.1^2)(640{,}000 + 1{,}000^2) - 0.1^2 1{,}000^2$
            = $10{,}500.$

5.80 If $\rho = 0$ then the claims from successive years are uncorrelated and hence the past data $\mathbf{x} = (X_1, \dots, X_n)$ are of no value in helping to predict X_{n+1} so more reliance should be placed on μ. (Unlike most models in this chapter, here we get to know μ as opposed to only knowing a probability distribution concerning μ.) Conversely, if $\rho = 1$, then X_{n+1} is a perfect linear function of $\mathbf{x}$. Thus no reliance need be placed on μ.

5.81 (a) $\dfrac{\Gamma'(\alpha)}{\Gamma(\alpha)} = \dfrac{\frac{d}{d\alpha} \int_0^\infty x^{\alpha-1} e^{-x} dx}{\Gamma(\alpha)} = \dfrac{\int_0^\infty (\log x) x^{\alpha-1} e^{-x} dx}{\Gamma(\alpha)}$

(b) $\Psi'(\alpha)$ = $\dfrac{d}{d\alpha} \dfrac{\Gamma'(\alpha)}{\Gamma(\alpha)} = \dfrac{\Gamma(\alpha)\Gamma''(\alpha) - [\Gamma'(\alpha)]^2}{[\Gamma(\alpha)]^2}$

                = $\dfrac{\Gamma''(\alpha)}{\Gamma(\alpha)} - [\Psi(\alpha)]^2$

                = $\dfrac{\int_0^\infty (\log x)^2 x^{\alpha-1} e^{-x} dx}{\Gamma(\alpha)} - [\Psi(\alpha)]^2$

$$\int_0^\infty (\log x)^2 x^{\alpha-1} e^{-x} dx = \Gamma(\alpha)\{\Psi'(\alpha) + [\Psi(\alpha)]^2\}$$

5.82 The Bayes estimator is found from the posterior distribution,

$$\pi_{\Theta|\mathbf{X}}(\theta|\mathbf{x}) \propto \exp\left[-\frac{1}{2}\sum_{j=1}^{25}\left(\frac{\log x_j - \theta}{2}\right)^2 - \frac{1}{2}(\theta - 5)^2 \right]$$

$$\propto \ \exp\left\{-\frac{1}{2}\left[\frac{\theta-\frac{2}{29}(10+\frac{1}{2}\sum\log x_j)}{\sqrt{4/29}}\right]^2\right\}$$

which implies that the posterior distribution of Θ is normal with mean $\frac{2}{29}(10+\frac{1}{2}\sum\log x_j)$ and variance $4/29$. We are trying to estimate the mean of X and with a lognormal distribution, it is $\exp(\theta+2)$. The Bayes estimator is the posterior expected value of this quantity which is e^2 times the moment generating function of a normal random variable evaluated at 1 (recall that the normal *mgf* is $M(t)=\exp(\mu t+\sigma^2 t^2/2)$ and so

$$\hat{\mu}_{Bayes} \ = \ \exp\left[2+\frac{2}{29}(10+\frac{1}{2}\sum\log x_j)+\frac{2}{29}\right]$$

$$= \ \exp\left(\frac{80}{29}+\frac{25}{29}\bar{W}\right)$$

where $\bar{W}=\frac{1}{25}\sum\log x_j$.

The credibility estimator is found from

$$\mu(\Theta) \ = \ e^{\Theta+2}$$
$$\mu \ = \ E(e^{\Theta+2})=e^{2+5+0.5}=e^{7.5}$$
$$v(\Theta) \ = \ e^{2\Theta+8}-e^{2\Theta+4}$$
$$v \ = \ E(e^{2\Theta+8}-e^{2\Theta+4})=e^{10+2+8}-e^{10+2+4}=e^{16}(e^4-1)$$
$$a \ = \ E(e^{2\Theta+4})-e^{15}=e^{16}-e^{15}=e^{16}(e-1)$$
$$Z \ = \ \frac{25}{25+\frac{e^4-1}{e-1}}=.44490$$
$$\hat{\mu}_{cred} \ = \ .44490\bar{X}+.55510e^{7.5}.$$

The log-credibility estimator is found from

$$\mu(\Theta) \ = \ E(\log X|\Theta)=\Theta$$
$$\mu \ = \ E(\Theta)=5$$
$$v(\Theta) \ = \ Var(\log X|\Theta)=4$$
$$v \ = \ E(4)=4$$
$$a \ = \ Var(\Theta)=1$$
$$Z \ = \ \frac{25}{25+\frac{4}{1}}=\frac{25}{29}$$
$$\hat{\mu}_{\log-cred} \ = \ c\exp\left(\frac{25}{29}\bar{W}+\frac{4}{29}5\right).$$

We know that the Bayes estimator is unbiased and because the log-credibility and Bayes estimators both involve $\exp(25\bar{W}/29)$ and both are unbiased, they must be identical. Therefore

$$\hat{\mu}_{\log-cred}=\exp\left(\frac{25}{29}\bar{W}+\frac{80}{29}\right).$$

With regard to bias and mean squared error, for the Bayes and log-credibility estimators,

$$
\begin{aligned}
E(\hat{\mu}|\theta) &= e^{80/29}E(e^{25\bar{W}/29}|\theta) \\
&= e^{80/29}[E(X^{1/29}|\theta)]^{25} \\
&= e^{80/29}\left\{\exp\left[\frac{1}{29}\theta + \frac{1}{2}\left(\frac{1}{29}\right)^2 4\right]\right\}^{25} \\
&= \exp\left(\frac{25}{29}\theta + \frac{2370}{841}\right) \\
bias &= \exp\left(\frac{25}{29}\theta + \frac{2370}{841}\right) - \exp(\theta + 2) \\
E(\hat{\mu}^2|\theta) &= e^{160/29}E(e^{50\bar{W}/29}|\theta) \\
&= e^{160/29}[E(X^{2/29}|\theta)]^{25} \\
&= e^{160/29}\left\{\exp\left[\frac{2}{29}\theta + \frac{1}{2}\left(\frac{2}{29}\right)^2 4\right]\right\}^{25} \\
&= \exp\left(\frac{50}{29}\theta + \frac{4840}{841}\right) \\
variance &= \exp\left(\frac{50}{29}\theta + \frac{4840}{841}\right) - \exp\left(\frac{50}{29}\theta + \frac{4740}{841}\right) \\
mse &= varaiance + bias^2
\end{aligned}
$$

For the credibility estimator

$$
\begin{aligned}
E(\hat{\mu}|\theta) &= E(0.4449\bar{X} + 0.5551e^{7.5}) \\
&= 0.4449e^{\theta+2} + 0.5551e^{7.5} \\
bias &= 0.5551(e^{7.5} - e^{\theta+2}) \\
Var(\hat{\mu}|\theta) &= 0.4449^2 Var(\bar{X}) \\
&= 0.4449^2(e^{2\theta+8} - e^{2\theta+4})/25
\end{aligned}
$$

Values of the bias and mean squared error are given for various percentiles in Table 4.10.

5.83 (a)

$$
\begin{aligned}
E\{[X_{n+1} - g(\mathbf{X})]^2\} &= E\{[X_{n+1} - E(X_{n+1}|\mathbf{X}) + E(X_{n+1}|\mathbf{X}) - g(\mathbf{X})]^2\} \\
&= E\{[X_{n+1} - E(X_{n+1}|\mathbf{X})]^2\} \\
&\quad + E\{[E(X_{n+1}|\mathbf{X}) - g(\mathbf{X})]^2\} \\
&\quad + 2E\{[X_{n+1} - E(X_{n+1}|\mathbf{X})][E(X_{n+1}|\mathbf{X}) - g(\mathbf{X})]\}
\end{aligned}
$$

The third term is

$$
2E\{[X_{n+1} - E(X_{n+1}|\mathbf{X})][E(X_{n+1}|\mathbf{X}) - g(\mathbf{X})]\}
$$

Table 4.10   Results for Exercise 5.82

Percentile	θ	mean	Bayes/log-cred bias	mse	credibility bias	mse
1	2.674	107	61	6,943	944	893,379
5	3.355	212	90	18,691	886	804,248
10	3.718	304	109	31,512	835	735,976
25	4.326	559	138	75,310	694	613,435
50	5.000	1,097	150	202,165	395	666,286
75	5.674	2,153	78	580,544	-191	2,003,192
90	6.282	3,950	-186	1,671,092	-1,189	8,036,278
95	6.645	5,681	-533	3,345,035	-2,150	18,316,467
99	7.326	11,230	-1,966	13,777,997	-5,230	80,871,171

$$= 2E(E\{[X_{n+1} - E(X_{n+1}|\mathbf{X})][E(X_{n+1}|\mathbf{X}) - g(\mathbf{X})]|\mathbf{X}\})$$
$$= 2E\{[E(X_{n+1}|\mathbf{X}) - E(X_{n+1}|\mathbf{X})][E(X_{n+1}|\mathbf{X}) - g(\mathbf{X})]\}$$
$$= 0$$

completing the proof.

(b) The objective function is minimized when $E\{[E(X_{n+1}|\mathbf{X}) - g(\mathbf{X})]^2\}$ is minimized. If $g(\mathbf{X})$ is set equal to $E(X_{n+1}|\mathbf{X})$ the expectation is of a random variable which is identically zero and so is zero. Because an expected square cannot be negative, this is the minimum. But this is the Bayesian premium.

(c) Inserting a linear function, the mean squared error to minimize is

$$E\{[E(X_{n+1}|\mathbf{X}) - \alpha_0 - \sum_{j=1}^{n} \alpha_j X_j]^2\}.$$

But this is (5.54) which is minimized by the linear credibility premium.

4.4   SECTION 5.5

5.84 $\bar{X}_1 = 733\frac{1}{3}$, $\bar{X}_2 = 633\frac{1}{3}$, $\bar{X}_3 = 900$, $\bar{X} = 755\frac{5}{9}$

$\hat{v}_1 = (16\frac{2}{3}^2 + 66\frac{2}{3}^2 + 83\frac{1}{3}^2)/2 = 5,833\frac{1}{3}$

$\hat{v}_2 = (8\frac{1}{3}^2 + 33\frac{1}{3}^2 + 41\frac{2}{3}^2)/2 = 1,458\frac{1}{3}$

$\hat{v}_3 = (0^2 + 50^2 + 50^2)/2 = 2,500$

$\hat{v} = 3,263\frac{8}{9}$

$\hat{a} = (22\frac{2}{9}^2 + 122\frac{2}{9}^2 + 144\frac{4}{9}^2) - 3,263\frac{8}{9}^2/3 = 17,060\frac{5}{27}$

$$\hat{k} = 3,263\frac{8}{9}/17,060\frac{5}{27} = 0.191316$$
$$Z = 3/(3 + 0.191316) = 0.94005$$

The three estimates are

$$0.94005(733\frac{1}{3}) + 0.05995(755\frac{5}{9}) = 734.67$$
$$0.94005(633\frac{1}{3}) + 0.05995(755\frac{5}{9}) = 640.66$$
$$0.94005(900) + 0.05995(755\frac{5}{9}) = 891.34.$$

5.85
$$\bar{X}_1 = 45,000/220 = 204.55$$
$$\bar{X}_2 = 54,000/235 = 229.79$$
$$\bar{X}_3 = 91,000/505 = 180.20$$
$$\hat{\mu} = \bar{X} = 190,000/960 = 197.91.$$

$$
\begin{aligned}
\hat{v} &= [100(4.55)^2 + 120(3.78)^2 + 90(18.68)^2 + 75(10.21)^2 + 70(13.07)^2 \\
&\quad +150(6.87)^2 + 175(8.77)^2 + 180(14.24)^2/(1 + 2 + 2) \\
&= 19,651.07,
\end{aligned}
$$

$$\hat{a} = \frac{\begin{array}{c}220(204.55 - 197.91)^2 + 235(229.79 - 197.91)^2 \\ +505(180.20 - 197.91)^2 - 19,651.07(2)\end{array}}{960 - (220^2 + 235^2 + 505^2)/960} = 633.94.$$

$\hat{k} = 31.00$, $Z_1 = 220/251 = 0.8765$, $Z_2 = 235/266 = 0.8835$, $Z_3 = 505/536 = 0.9422$. The estimates are

$$0.8765(204.55) + 0.1235(197.91) = 203.73$$
$$0.8835(229.79) + 0.1165(197.91) = 226.08$$
$$0.9422(180.20) + 0.0578(197.91) = 181.22.$$

Using the alternative method,

$$\hat{\mu} = \frac{0.8765(204.55) + 0.8835(229.79) + 0.9422(180.20)}{0.8765 + 0.8835 + 0.9422} = 204.31$$

and the estimates are

$$0.8765(204.55) + 0.1235(204.31) = 204.52$$
$$0.8835(229.79) + 0.1165(204.31) = 226.82$$
$$0.9422(180.20) + 0.0578(204.31) = 181.59.$$

5.86 $\bar{X} = 475$, $\hat{v} = (0^2 + 75^2 + 75^2)/2 = 5{,}625$. With μ known to be 600, $\tilde{a} = (475 - 600)^2 - 5{,}625/3 = 13{,}750$, $\hat{k} = 5{,}625/1{,}375 = 0.4091$, $Z = 2/2.4091 = 0.8302$. The premium is $0.8302(475) + 0.1698(600) = 496.23$.

5.87 (a)
$$
\begin{aligned}
Var(X_{ij}) &= E[Var(X_{ij}|\Theta_i)] + Var[E(X_{ij}|\Theta_i)] \\
&= E[v(\Theta_i] + Var[\mu(\Theta_i)] = v + a.
\end{aligned}
$$

(b) This follows from (5.13).

(c)

$$
\begin{aligned}
\sum_{i=1}^{r}\sum_{j=1}^{n}(X_{ij} - \bar{X})^2 &= \sum_{i=1}^{r}\sum_{j=1}^{n}(X_{ij} - \bar{X}_i + \bar{X}_i - \bar{X})^2 \\
&= \sum_{i=1}^{r}\sum_{j=1}^{n}(X_{ij} - \bar{X}_i)^2 + 2\sum_{i=1}^{r}\sum_{j=1}^{n}(X_{ij} - \bar{X}_i)(\bar{X}_i - \bar{X}) \\
&\quad + \sum_{i=1}^{r}\sum_{j=1}^{n}(\bar{X}_i - \bar{X})^2 \\
&= \sum_{i=1}^{r}\sum_{j=1}^{n}(X_{ij} - \bar{X}_i)^2 + 2\sum_{i=1}^{r}(\bar{X}_i - \bar{X})\sum_{j=1}^{n}(X_{ij} - \bar{X}_i) \\
&\quad + n\sum_{i=1}^{r}(\bar{X}_i - \bar{X})^2 \\
&= \sum_{i=1}^{r}\sum_{j=1}^{n}(X_{ij} - \bar{X}_i)^2 + n\sum_{i=1}^{r}(\bar{X}_i - \bar{X})^2
\end{aligned}
$$

The middle term is zero because $\sum_{j=1}^{n}(X_{ij} - \bar{X}_i) = \sum_{j=1}^{n}X_{ij} - n\bar{X}_i = 0$.

(d)

$$
E\left[\frac{1}{nr-1}\sum_{i=1}^{r}\sum_{j=1}^{n}(X_{ij} - \bar{X})^2\right]
$$

$$
= E\left[\frac{1}{nr-1}\sum_{i=1}^{r}\sum_{j=1}^{n}(X_{ij} - \bar{X}_i)^2 + \frac{n}{nr-1}\sum_{i=1}^{r}(\bar{X}_i - \bar{X})^2\right]
$$

We know that $\frac{1}{n-1}\sum_{j=1}^{n}(X_{ij} - \bar{X}_i)^2$ is an unbiased estimator of $v(\theta_i)$ and so the expected value of the first term is

$$
E\left[\frac{1}{nr-1}\sum_{i=1}^{r}\sum_{j=1}^{n}(X_{ij} - \bar{X}_i)^2\right]
$$

$$
= E\left\{E\left[\frac{1}{nr-1}\sum_{i=1}^{r}\sum_{j=1}^{n}(X_{ij} - \bar{X}_i)^2|\Theta_i\right]\right\}
$$

$$= E\left[\frac{n-1}{nr-1}\sum_{i=1}^{r} v(\Theta_i)\right] = \frac{r(n-1)}{nr-1}v.$$

For the second term note that $Var(\bar{X}_i) = E[v(\Theta_i)/n] + Var[\mu(\Theta_i)] = v/n + a$ and $\frac{1}{r-1}\sum_{i=1}^{r}(\bar{X}_i - \bar{X})^2$ is an unbiased estimator of $v/n + a$. Then, for the second term,

$$E\left[\frac{n}{nr-1}\sum_{i=1}^{r}(\bar{X}_i - \bar{X})^2\right] = \frac{n(r-1)}{nr-1}\left(\frac{v}{n} + a\right).$$

Then the total is

$$\frac{r(n-1)}{nr-1}v + \frac{n(r-1)}{nr-1}\left(\frac{v}{n} + a\right) = v + a - \frac{n-1}{nr-1}a.$$

(e) Unconditionally, all of the X_{ij} are assumed to have the same mean, when in fact they do not. They also have a conditional variance which is smaller and so the variance from $\bar{X}$ is not as great as it appears from (b).

5.88 $\bar{X} = 333/2,787 = 0.11948 = \hat{v}$. The sample variance is $447/2,787 - 0.11948^2 = 0.14611$ and so $\hat{a} = 0.14611 - 0.11948 = 0.02663$. Then, $\hat{k} = 0.11948/0.02663 = 4.4867$ and $Z = 1/5.4867 = 0.18226$. The premiums are given in Table 4.11.

Table 4.11 Calculations for Exercise 5.88

No. of claims	Premium
0	$0.18226(0) + 0.81774(0.11948) = 0.09770$
1	$0.18226(1) + 0.81774(0.11948) = 0.27996$
2	$0.18226(2) + 0.81774(0.11948) = 0.46222$
3	$0.18226(3) + 0.81774(0.11948) = 0.64448$
4	$0.18226(4) + 0.81774(0.11948) = 0.82674$

5.89 (a) See Appendix B.

(b)
$$a = Var[\mu(\Theta)] = Var(\Theta)$$
$$v = E[v(\Theta)] = E[\Theta(1 + \Theta)]$$
$$\mu = E[\mu(\Theta)] = E(\Theta)$$
$$v - \mu - \mu^2 = E(\Theta) - E(\Theta^2) - E(\Theta) - E(\Theta)^2 = Var(\Theta) = a.$$

(c)
$$\hat{\mu} = \bar{X} = 0.11948$$
$$\hat{a} + \hat{v} = 0.14611$$
$$\hat{a} = \hat{v} - 0.11948 - 0.11948^2$$
$$\hat{a} - \hat{v} = -0.133755$$

The solution is $\hat{a} = 0.0061775$ and $\hat{v} = 0.1399325$. Then

$$\hat{k} = 0.1399325/0.0061775 = 22.652$$

and $Z = 1/23.652 = 0.04228$. The premiums are given in Table 4.12

Table 4.12 Calculations for Exercise 5.89

No. of claims	Premium
0	$0.04228(0) + 0.95772(0.11948) = 0.11443$
1	$0.04228(1) + 0.95772(0.11948) = 0.15671$
2	$0.04228(2) + 0.95772(0.11948) = 0.19899$
3	$0.04228(3) + 0.95772(0.11948) = 0.24127$
4	$0.04228(4) + 0.95772(0.11948) = 0.28355$

5.90

$$
\begin{aligned}
f_{\mathbf{X}_i}(\mathbf{x}_i) &= \int_0^\infty \prod_{j=1}^{n_i} \left[\frac{(m_{ij}\theta_i)^{t_{ij}} e^{-m_{ij}\theta_i}}{t_{ij}!} \right] \frac{1}{\mu} e^{-\theta_i/\mu} d\theta_i \\
&= \frac{1}{\mu} \left(\prod_{j=1}^{n_i} \frac{m_{ij}^{t_{ij}}}{t_{ij}!} \right) \int_0^\infty e^{-\theta_i(\mu^{-1}+m_i)} \theta_i^{t_i} d\theta_i \\
&= \frac{1}{\mu} \left(\prod_{j=1}^{n_i} \frac{m_{ij}^{t_{ij}}}{t_{ij}!} \right) \frac{t_i!}{(\mu^{-1}+m_i)^{t_i+1}} \propto \mu^{-1}(\mu^{-1} + m_i)^{-t_i-1}
\end{aligned}
$$

where $t_i = \sum_{j=1}^{n_i} t_{ij}$. Then the likelihood function is

$$L(\mu) \propto \mu^{-r} \prod_{i=1}^r (\mu^{-1} + m_i)^{-t_i-1}$$

and the logarithm is

$$l(\mu) = -r \log(\mu) - \sum_{i=1}^r (t_i + 1) \log(\mu^{-1} + m_i)$$

and

$$l'(\mu) = -r\mu^{-1} - \sum_{i=1}^r (t_i + 1)(\mu^{-1} + m_i)^{-1}(-\mu^{-2}) = 0.$$

The equation to be solved is

$$r\mu = \sum_{i=1}^r \frac{t_i + 1}{\mu^{-1} + m_i}.$$

5.91 (a)

$$
\sum_{i=1}^{r}\sum_{j=1}^{n_i} m_{ij}(X_{ij} - \bar{X})^2 = \sum_{i=1}^{r}\sum_{j=1}^{n_i} m_{ij}(X_{ij} - \bar{X}_i + \bar{X}_i - \bar{X})^2
$$

$$
= \sum_{i=1}^{r}\sum_{j=1}^{n_i} m_{ij}(X_{ij} - \bar{X}_i)^2
$$

$$
+ 2\sum_{i=1}^{r}\sum_{j=1}^{n_i} m_{ij}(X_{ij} - \bar{X}_i)(\bar{X}_i - \bar{X})
$$

$$
+ \sum_{i=1}^{r}\sum_{j=1}^{n_i} m_{ij}(\bar{X}_i - \bar{X})^2
$$

$$
= \sum_{i=1}^{r}\sum_{j=1}^{n_i} m_{ij}(X_{ij} - \bar{X}_i)^2 + \sum_{i=1}^{r} m_i(\bar{X}_i - \bar{X})^2.
$$

The middle term vanishes because $\sum_{j=1}^{n_i} m_{ij}(X_{ij} - \bar{X}_i) = 0$ from the definition of $\bar{X}_i$.

(b)

$$
\hat{a} = (m - m^{-1}\sum_{i=1}^{r} m_i^2)^{-1}\left[\sum_{i=1}^{r}\sum_{j=1}^{n_i} m_{ij}(X_{ij} - \bar{X})^2\right.
$$

$$
\left. - \sum_{i=1}^{r}\sum_{j=1}^{n_i} m_{ij}(X_{ij} - \bar{X}_i)^2 - (r - 1)\hat{v}\right]
$$

$$
= (m - m^{-1}\sum_{i=1}^{r} m_i^2)\left[-1\sum_{i=1}^{r}\sum_{j=1}^{n_i} m_{ij}(X_{ij} - \bar{X})^2\right.
$$

$$
\left. - \hat{v}\sum_{i=1}^{r}(n_i - 1) - (r - 1)\hat{v}\right].
$$

Also

$$
m_* = \frac{\sum_{i=1}^{r} m_i\left(1 - \frac{m_i}{m}\right)}{\sum_{i=1}^{r} n_i - 1} = \frac{m - m^{-1}\sum_{i=1}^{r} m_i^2}{\sum_{i=1}^{r} n_i - 1}.
$$

Then

$$
\hat{a} = \frac{m_*^{-1}}{\sum_{i=1}^{r} n_i - 1}\left[\sum_{i=1}^{r}\sum_{j=1}^{n_i} m_{ij}(X_{ij} - \bar{X})^2 - \hat{v}\left(\sum_{i=1}^{r} n_i - 1\right)\right]
$$

$$
= m_*^{-1}\left[\frac{\sum_{i=1}^{r}\sum_{j=1}^{n_i} m_{ij}(X_{ij} - \bar{X})^2}{\sum_{i=1}^{r} n_i - 1} - \hat{v}\right].
$$

5.92 The sample mean is $21/34$ and the sample variance is $370/340 - (21/34)^2 = 817/1{,}156$. Then $\hat{v} = 21/34$ and $\hat{a} = 817/1{,}156 - 21/34 = 103/1{,}156$.

$\hat{k} = (21/34)/(103/1{,}156) = 714/103$ and $Z = 1/(1 + 714/103) = 103/817$. The estimate is $(103/817)(2) + (714/817)(21/34) = 0.79192$.

Chapter 6 Solutions

5.1 SECTION 6.3

6.1 (a) $\tilde{\psi}(2,0) = 0, \quad f_1 = \Pr(U_0^* = 2) = 1$

k	S_k	$w_{1,k} = 6 + (6+2) \times 10\% - S_k$	$u_0 + w_{1,k}$	$g_{1,k}$
1	0	6.8	8.8	0.40
2	5	1.8	3.8	0.30
3	10	−3.2	−1.2	0.15
4	15	−8.2	−6.2	0.10
5	20	−13.2	−11.2	0.05

$$\tilde{\psi}(2,1) = \Pr(w_{q,k} + 2 < 0) = 0.15 + 0.1 + 0.05 = 0.3.$$

The updating equation is $u_j + w_{j,k} = (u_j + 6) \times 1.1 - s_k$. The next calculations are in Table 5.1. Then,

$$\tilde{\psi}(2,2) = \tilde{\psi}(2,1) + 0.3(0.1) + 0.3(0.05) + 0.4(0.05) = 0.365.$$

The next calculations are also in Table 5.1. Then,

$$
\begin{aligned}
\tilde{\psi}(2,3) &= \tilde{\psi}(2,2) + 0.15(0.045 + 0.04) + 0.1(-0.045 + 0.04 + 0.09 + 0.06) \\
&\quad + 0.05(0.045 + 0.04 + 0.09 + 0.06 + 0.12 + 0.12) \\
&= 0.365 + 0.06 = 0.425
\end{aligned}
$$

Table 5.1 Calculations for Exercise 6.1, part (a)

Calculations for year 2

| | | | | | $u_j + w_{j,k}; g_{j,k}$ | | |
| | | | | | k | | |
j	$U_1^* = u_j$	f_j	1	2	3	4	5
			0.4	0.3	0.15	0.1	0.05
1	3.8	0.3	10.78	5.78	0.78	−4.22	−9.22
2	8.8	0.4	16.28	11.28	6.28	1.28	−3.72
			0.4	0.3	0.15	0.1	0.05

Calculations for year 3

| | | | | | $u_j + w_{j,k}; g_{j,k}$ | | |
| | | | | | k | | |
j	$U_2^* = u_j$	f_j	1	2	3	4	5
			0.4	0.3	0.15	0.1	0.05
1	0.78	0.045	7.458	2.458	−2.542	−7.542	−12.542
2	1.28	0.04	8.008	3.008	−1.992	−6.992	−11.992
3	5.78	0.09	12.958	7.958	2.958	−2.042	−7.042
4	6.28	0.06	13.508	8.508	3.508	−1.492	−6.492
5	10.78	0.12	18.458	13.458	8.458	3.458	−1.542
6	11.28	0.12	19.008	14.008	9.008	4.008	−0.992
7	16.28	0.16	24.508	19.508	14.508	9.508	4.508
			0.4	0.3	0.15	0.1	0.05

(b)

j	$U_1^* = u_j$	f_j
1	3.8	0.3
2	8.8	0.4

The next set of calculations appears in the first half of Table 5.3. Then,

$$
\begin{aligned}
\tilde{\psi}(2,2) &= \tilde{\psi}(2,1) + 0.072(0.15 + 0.1 + 0.05) + 0.324(0.1 + 0.05) \\
&\quad + 0.304(0.05) \\
&= 0.3 + 0.0854 = 0.3854
\end{aligned}
$$

Values at the end of year two are in Table 5.2.

Table 5.2   Calculations for Exercise 6.1 part (b), year 2

j	$U_2^* = u_j$	f_j
1	1.6	0.0216
2	2.1	0.0486
3	2.6	0.0304
4	6.6	0.0288
5	7.1	0.0972
6	7.6	0.0456
7	12.1	0.1296
8	12.6	0.0912
9	17.6	0.1216

The next set of calculations appears in the second half of Table 5.3. Then,

$$
\begin{aligned}
\tilde{\psi}(2,3) &= \tilde{\psi}(2,2) + 0.057468(0.15 + 0.1 + 0.05) + 0.140980(0.1 + 0.05) \\
&\quad + 0.192696(0.05) \\
&= 0.3854 + 0.0480222 = 0.4334222 > 0.425.
\end{aligned}
$$

Table 5.3 Calculations for Exercise 6.1, part (b)

Calculations for year 2

j	$U_1^* = u_j$	f_j	\multicolumn 1		2		3		4		5	
			\multicolumn{10}{}{$u_j + w_{j,k};\, g_{j,k}$}									
			\multicolumn{10}{}{k}									
1	0	0.072	6.6	0.4	1.6	0.3	−3.4	0.15	−8.4	0.10	−13.4	0.05
2	5	0.324	12.1	0.4	7.1	0.3	2.1	0.15	−2.9	0.10	−7.9	0.05
3	10	0.304	17.6	0.4	12.6	0.3	7.6	0.15	2.6	0.10	−2.4	0.05

Calculations for year 3

j	$U_2^* = u_j$	f_j	1		2		3		4		5	
			\multicolumn{10}{}{$u_j + w_{j,k};\, g_{j,k}$}									
			\multicolumn{10}{}{k}									
1	0	0.057468	6.6	0.4	1.6	0.3	3.4	0.15	−8.4	0.1	−13.4	−.05
2	5	0.140980	12.1	0.4	7.1	0.3	2.1	0.15	−2.9	0.1	−7.9	0.05
3	10	0.192696	17.6	0.4	12.6	0.3	7.6	0.15	2.6	0.1	−2.4	0.05
4	15	0.160224	23.1	0.4	18.1	0.3	13.1	0.15	8.1	0.1	3.1	0.05
5	20	0.063232	28.6	0.4	23.6	0.3	18.6	0.15	13.6	0.1	8.6	0.05

From (a) we get the calculations in Table 5.4.

Table 5.4 Calculations for Exercise 6.1 year 3

j	$U_3^* = u_j$	f_j	j	$U_3^* = u_j$	f_j
1	1.6	0.0172404	11	12.6	0.0578088
2	2.1	0.0211470	12	13.1	0.0240336
3	2.6	0.0192696	13	13.6	0.0063232
4	3.1	0.0080112	14	17.6	0.0770784
5	6.6	0.0229872	15	18.1	0.0480672
6	7.1	0.0422940	16	18.6	0.0094848
7	7.6	0.0289044	17	23.1	0.0640896
8	8.1	0.0160224	18	23.6	0.0189696
9	8.6	0.0031616	19	28.6	0.0252928
10	12.1	0.0563920			

which, upon rounding, leads to Table 5.5.

Table 5.5 Calculations for Exercise 6.1 year 3

j	$U_3^* = u_j$	f_j
1	0	0.0362824
2	5	0.0903955
3	10	0.1237188
4	15	0.1311177
5	20	0.1063770
6	25	0.0604756
7	30	0.0182108

6.2 The initial calculations are in Table 5.6.

$$\tilde{\psi}(2,1) = \tilde{\psi}(2,0) + (0.15 + 0.1 + 0.05)[(1 - \tilde{\psi}(2,0)] = 0.3$$

Calculations for the next year appear in Tables 5.7–5.10.
The Fourier calculations appear in Table 5.13.

$$\tilde{\psi}(2,2) = \tilde{\psi}(2,1) + (0.00239 + 0.01933 + 0.06077)[1 - \tilde{\psi}(2,1)] = 0.35774.$$

The next set of calculations appear in Tables 5.11 and 5.12.
The Fourier calculations appear in Table 5.14.

Table 5.6   Calculations for Exercise 6.2

k	s_k	$w = 6(1 + 10\%) - s_k$	$\Pr(W = w)$	$u_1 = u_0(1 + 10\%) + w$
1	0	6.6	0.4	8.8
2	5	1.6	0.3	3.8
3	10	−3.4	0.15	−1.2
4	15	−8.4	0.1	−6.2
5	20	−13.4	0.05	−11.2

Table 5.7   Calculations for Exercise 6.2

j	$u^* = u \times 1.1$	$\Pr(U_1^{**} = u^*)$
1	4.18	3/7
2	9.68	4/7

Table 5.8   Calculations for Exercise 6.2

j	$u^* = u \times 1.1$	$\Pr(U_1^{**} = u^*)$
1	0	0.492/7
2	5	2.764/7
3	10	3.744/7

Table 5.9   Calculations for Exercise 6.2

w	$\Pr(W = w)$
6.6	0.40
1.6	0.30
−2.4	0.15
−8.4	0.10
−13.4	0.05

Table 5.10   Calculations for Exercise 6.2

w	$w + 15$	$\Pr(W = w + 15)$
−15	0	0.034
−10	5	0.084
−5	10	0.134
0	15	0.252
5	20	0.368
10	25	0.128

Table 5.11   Calculations for Exercise 6.2

j	u_2	$u^* = u_2 \times 1.1$	$\Pr\left(U_2^{**} = u^*\right)$
1	0	0	0.12594
2	5	5.5	0.21476
3	10	11.0	0.31508
4	15	16.5	0.26961
5	20	22.0	0.07462

Table 5.12   Calculations for Exercise 6.2

j	u^*	$\Pr\left(U_2^{**} = u^*\right)$
1	0	0.12594
2	5	0.195714
3	10	0.271110
4	15	0.251743
5	20	0.125665
6	25	0.029848

Table 5.13 Fourier calculations for Exercise 6.2

u	$f_1^{**}(u)$	$f_W(u)$	$\varphi_{1,2}$	$\varphi_{2,2}$	$\varphi_{3,2} = \varphi_{1,2} \times \varphi_{2,2}$	$f_2(u)$	$U_2 = u$	$\Pr(U_2^* = u)$
0	$\frac{0.492}{7}$	0.034	1	1	1	0.00239	−15	—
5	$\frac{2.764}{7}$	0.084	$0.3495 - 0.8141i$	$-0.5433 - 0.2811i$	$-0.4187 + 0.3440i$	0.01933	−10	—
10	$\frac{3.744}{7}$	0.134	$-0.4646 - 0.3945i$	$0.2680 + 0.0400i$	$-0.1087 - 0.1244i$	0.06077	−5	—
15	0	0.252	$-0.2089 + 0.2557i$	$-0.1247 - 0.0131i$	$0.0294 - 0.0291i$	0.11555	0	0.12594
20	0	0.368	0.2103	0.072	0.0151	0.19704	5	0.21476
25	0	0.128	$-0.2089 - 0.2557i$	$-0.1247 + 0.0131i$	$0.0294 + 0.0291i$	0.28909	10	0.31508
30	0	0	$-0.4646 + 0.3949i$	$0.2680 - 0.0400i$	$-0.1087 + 0.1244i$	0.24737	15	0.26961
35	0	0	$0.3495 + 0.8141i$	$-0.5433 + 0.2811i$	$-0.4187 - 0.3440i$	0.06846	20	0.07462

Table 5.14 Fourier calculations for Exercise 6.2

u	$f_2^{**}(u)$	$f_W(u)$	$\varphi_{1,3}$	$\varphi_{2,3}$	$\varphi_{3,3} = \varphi_{1,3} \times \varphi_{2,3}$	$f_3(u)$	$U_3 = u$
0	0.12594	0.034	1	1	1	0.00428	−15
5	0.195714	0.084	$0.58338 - 0.65242i$	$0.25381 - 0.84597i$	$-0.40386 - 0.65911i$	0.01723	−10
10	0.271110	0.134	$-0.06045 - 0.56640i$	$-0.54330 - 0.28108i$	$-0.12636 + 0.32472i$	0.04253	−5
15	0.251743	0.252	$-0.19587 - 0.13909i$	$-0.14317 + 0.34106i$	$0.07548 - 0.04689i$	0.08930	0
20	0.125655	0.368	$-0.01951 + 0.02618i$	$0.26800 + 0.04000i$	$-0.00627 + 0.00624i$	0.15741	5
25	0.029848	0.128	$0.06434 - 0.00702i$	$0.02166 - 0.20543i$	$-0.00005 - 0.01337i$	0.20177	10
30	0	0	$0.06100 - 0.02418i$	$-0.12470 - 0.01308i$	$-0.00792 + 0.00222i$	0.20761	15
35	0	0	$0.05191 - 0.01768i$	$0.00369 + 0.07953i$	$0.00160 + 0.00406i$	0.16301	20
40	0	0	0.04541	0.072	0.00327	0.08599	25
45	0	0	$0.05191 + 0.01768i$	$0.00369 - 0.07953i$	$0.00160 - 0.00406i$	0.02707	30
50	0	0	$0.06100 + 0.02418i$	$-0.12470 + 0.01308i$	$-0.00792 - 0.00222i$	0.00382	35
55	0	0	$0.06434 + 0.00702i$	$0.02166 + 0.20543i$	$-0.00005 + 0.0133i$	0	40
60	0	0	$-0.01951 - 0.02618i$	$0.26800 - 0.04000i$	$-0.00627 - 0.00624i$	0	45
65	0	0	$-0.19587 + 0.13909i$	$-0.14317 - 0.34106i$	$0.07548 + 0.04689i$	0	50
70	0	0	$-0.06045 + 0.56640i$	$-0.54330 + 0.28108i$	$-0.12636 - 0.32472i$	0	55
75	0	0	$0.58338 + 0.65242i$	$0.25381 + 0.84597i$	$-0.40386 + 0.65911i$	0	60

5.2   SECTION 6.4

6.3
$$Pr\left(N_{t+s} - N_s = 1\right) \;\; = \;\; \lambda t \cdot e^{-\lambda t} = \lambda t \sum_{n=0}^{\infty} \frac{(-\lambda t)^n}{n!}$$

$$= \;\; \lambda t + \sum_{n=1}^{\infty} \lambda t \frac{(-\lambda t)^n}{n!} = \lambda t + o(t)$$

since $\lim\limits_{t \to 0} \frac{1}{t} \sum\limits_{n=1}^{\infty} \lambda t \frac{(-\lambda t)^n}{n!} = \lambda \lim\limits_{t \to 0} \frac{(-\lambda t)^n}{n!} = 0.$

6.4 We begin by assuming that there were i claims up to time t_0 and that the ith claim was at time t. Then, let T be the time to the next claim, so

$$\Pr(T \;\; > \;\; t_0 + s | N_{t_0} = i, i\text{th claim at time } t)$$
$$= \;\; \Pr(W_{i+1} > t_0 + s - t | N_{t_0} = i, W_{i+1} > t_0 - t)$$
$$= \;\; \frac{\Pr(W_{i+1} > t_0 + s - t, N_{t_0} = i)}{\Pr(W_{i+1} > t_0 - t, N_{t_0} = i)}$$
$$= \;\; \frac{\Pr(W_{i+1} > t_0 + s - t | N_{t_0} = i)\,\Pr(N_{t_0} = i)}{\Pr(W_{i+1} > t_0 - t | N_{t_0} = i)\,\Pr(N_{t_0} = i)}$$
$$= \;\; \frac{\Pr(W_{i+1} > t_0 + s - t | N_{t_0} = i)}{\Pr(W_{i+1} > t_0 - t | N_{t_0} = i))}$$
$$= \;\; \frac{\exp[-\lambda(t_0 + s - t)]}{\exp[-\lambda(t_0 - t)]}$$
$$= \;\; \exp(-\lambda s).$$

and since the answer does not depend on t or i, the result holds.

5.3   SECTION 6.5

6.5 κ is the smallest positive root of $1 + (1+\theta)2\beta\kappa = (1-\beta\kappa)^{-2}$, $\theta = 0.32 \Rightarrow$ $\beta\kappa(6\beta\kappa - 1)(11\beta\kappa - 16) = 0 \Rightarrow \kappa = \frac{1}{6\beta}$.

6.6 $f(x) \propto x^{-1/2}e^{-\beta x}$ and so X has the gamma distribution with $\alpha = 1/2$ and $\theta = 1/\beta$. Then $\mu = E(X) = \frac{1}{2\beta}$, $M_X(t) = E\left(e^{tX}\right) = \left(\frac{1}{1-t/\beta}\right)^{1/2}$, $t < \beta$. κ is the smallest postivie root of $1 + (1+\theta)\mu\kappa = E\left(e^{\kappa X}\right)$ and so $1 + (1+\theta)\frac{1}{2\beta}\kappa = (1 - \kappa/\beta)^{-1/2}$.

Let $y = \frac{\kappa}{\beta}$. Then $1 + (1+\theta)\frac{y}{2} = (1-y)^{-1/2}$. Square both sides to obtain $\left[1 + (1+\theta)\frac{y}{2}\right]^2 = (1-y)^{-1}$. Then multiply by $1-y$, divide by y, and multiply by 4 to obtain
$$(1+\theta)^2 y^2 + (1+\theta)(3-\theta)y - 4\theta = 0$$

The solutions to this quadratic equation are

$$y = \frac{-(1+\theta)(3-\theta) \pm \sqrt{[(1+\theta)(3-\theta)]^2 + 16\theta(1+\theta)^2}}{2(1+\theta)^2}$$

$$= \frac{(\theta-3) \pm \sqrt{(\theta+1)(\theta+9)}}{2(1+\theta)}$$

Because $\theta \geq 0$, $|\theta-3| < |\sqrt{(\theta+1)(\theta+3)}|$ and so the positive root is the only one which has a chance to yield a postive value for y. Becaue $(\theta+1)(\theta+9) = \theta^2 100 + 9 \geq 9$ when $\theta \geq 0$ a positive value for y is assured. Then we get $y = \frac{(\theta-3)+\sqrt{(\theta+1)(\theta+0)}}{2(1+\theta)}$ and so $\kappa = y\beta = \frac{(\theta-3)+\sqrt{(\theta+1)(\theta+9)}}{2(\theta+1)}\beta$

6.7

$$f(x) = \frac{1}{2}\left(2e^{-2x}\right) + \frac{1}{2}\left(3e^{-3x}\right)$$

$$E(X) = \frac{1}{2}\int_0^\infty 2xe^{-2x}dx + \frac{1}{2}\int_0^\infty 3xe^{-3x}dx = \frac{1}{2}\left(\frac{1}{2}+\frac{1}{3}\right) = \frac{5}{12}$$

$$3 = c = (1+\theta)\lambda\mu = (1+\theta)(4)\frac{5}{12} \Rightarrow \theta = \frac{4}{5}$$

$$E\left(e^{tX}\right) = \frac{1}{2}\left[2\int_0^\infty e^{-(2-t)x}dx + 3\int_0^\infty e^{-(3-t)x}dx\right]$$

$$= \frac{1}{2}\left(\frac{2}{2-t} + \frac{3}{3-t}\right), \, t < 2.$$

κ is the smallest positive root of $1 + (1+\theta)\mu\kappa = E\left(e^{\kappa X}\right)$. Therefore

$$1 + \left(1+\frac{4}{5}\right)\frac{5}{12}\kappa = E\left(e^{tX}\right) = \frac{1}{2}\left(\frac{2}{2-\kappa} + \frac{3}{3-\kappa}\right)$$

which implies $1+\frac{3}{4}\kappa = \frac{1}{2}\left(\frac{2}{2-\kappa} + \frac{3}{3-\kappa}\right)$. The roots are 0,1 and $\frac{8}{3}$ and so $\kappa = 1$.

6.8

$$\mu = E(X) = 1(0.2) + 2(0.3) + 3(0.5) = 2.3$$

$$E\left(X^2\right) = 1(0.2) + 4(0.3) + 9(0.5) = 5.9$$

$$c = (1+\theta)\mu\lambda \Rightarrow \theta\frac{c}{\mu\lambda} - 1 = \frac{2.99}{2.3} - 1 = 0.3$$

κ must be less than $\kappa_0 = 2\theta\mu/E\left(X^2\right) = 2(0.3)(2.3)/5.9 = 0.233898$. Let

$$H(t) = 1 + (1+\theta)\mu t - E\left(e^{tX}\right) = 1 + 2.99t - 0.2e^t - 0.3e^{2t} - 0.5e^{3t}$$

$$H'(t) = 2.99 - 0.2e^t - 0.6e^{2t} - 1.5e^{3t}$$

Use

$$\kappa_{n+1} = \kappa_n - \frac{H(\kappa_n)}{H'(\kappa_n)} = \kappa_n - \frac{1 + 2.99\kappa_n - 0.2e^{\kappa_n} - 0.3e^{2\kappa_n} - 0.5e^{3\kappa_n}}{2.99 - 0.2e^{\kappa_n} - 0.6e^{2\kappa_n} - 1.5e^{3\kappa_n}}.$$

With $\kappa_0 = 0.233898$, the iterations follow in Table 5.15 with the answer being $\kappa = 0.194273$.

Table 5.15 Iterations for the adjustment coefficient in Exercise 6.8

n	κ_n	
0	0.233898	
1	0.201104	
2	0.194539	
3	0.194274	
4	0.194273	$\leftarrow \kappa$ the adjustment coefficient

6.9 $f(x) = \frac{1}{2}\left[2e^{-2x} + 3e^{-3x}\right]$. From Exercise 6.7, $\theta = \frac{4}{5}, \mu = \frac{5}{12}$.

$$\begin{aligned}
E\left(X^2\right) &= \frac{1}{2}\left[2\int_0^\infty x^2 e^{-2x}\,dx + 3\int_0^\infty x^2 e^{-3x}\,dx\right] \\
&= \frac{1}{2}\left[2\Gamma(3)\left(\frac{1}{2}\right)^3 + 3\Gamma(3)\left(\frac{1}{3}\right)^3\right] = \frac{13}{36}.
\end{aligned}$$

κ must be less than $\kappa_0 = 2\theta\mu/E\left(X^2\right) = 2\frac{4}{5}\frac{5}{12}\frac{36}{13} = \frac{24}{13}$

$$\begin{aligned}
H(t) &= 1 + (1+\theta)\mu t - E\left(e^{tX}\right) = 1 + \frac{3}{4}t - \frac{1}{2}\left(\frac{2}{2-t} + \frac{3}{3-t}\right) \\
H'(t) &= \frac{3}{4} - \frac{1}{2}\left[\frac{2}{(2-t)^2} + \frac{3}{(3-t)^2}\right]
\end{aligned}$$

Use

$$\kappa_{n+1} = \kappa_n - \frac{H(\kappa_n)}{H'(\kappa_n)} = \kappa_n - \frac{1 + \frac{3}{4}\kappa_n - \frac{1}{2}\left(\frac{2}{2-\kappa_n} + \frac{3}{3-\kappa_n}\right)}{\frac{3}{4} - \frac{1}{2}\left[\frac{2}{(2-\kappa_n)^2} + \frac{3}{(3-\kappa_n)^2}\right]}.$$

With $\kappa_0 = \frac{24}{13}$ then the iterations are given in Table 5.16 with the solution being $\kappa = 1$.

6.10 κ satisfies

$$1 + (1+\theta)\mu\kappa = E\left(e^{\kappa X}\right) = E\left(1 + \kappa X + \frac{1}{2}\kappa^2 X^2 + \frac{1}{6}\kappa^3 X^3 + \cdots\right)$$

Table 5.16   Iterations for the adjustment coefficient in Exercise 6.9

n	κ_n	
0	1.846154	
1	1.719112	
2	1.528918	
3	1.305120	
4	1.117793	
5	1.021670	
6	1.000856	
7	1.000001	
8	1.000000	$\leftarrow \kappa$ the adjustment coefficient.

$$> \quad E\left(1 + \kappa X + \frac{1}{2}\kappa^2 X^2 + \frac{1}{6}\kappa^3 X^3\right)$$

$$= \quad 1 + \kappa\mu + \frac{1}{2}\kappa^2 E\left(X^2\right) + \frac{1}{6}\kappa^3 E\left(X^3\right)$$

This implies $E\left(X^3\right)\kappa^2 + 3E\left(X^2\right)\kappa - 6\theta\mu < 0$ and therefore

$$\kappa < \frac{-3E\left(X^2\right) + \sqrt{9[E\left(X^2\right)]^2 + 24\theta\mu E\left(X^3\right)}}{2E\left(X^3\right)}.$$

Then,

$$\frac{-3E\left(X^2\right) + \sqrt{9[E\left(X^2\right)]^2 + 24\theta\mu E\left(X^3\right)}}{2E\left(X^3\right)} < \frac{2\theta\mu}{E\left(X^2\right)}.$$

Since $X \geq 0$, the following are true

$$\sqrt{9[E\left(X^2\right)]^2 + 24\theta\mu E\left(X^3\right)} \quad < \quad \frac{2\theta\mu}{E\left(X^2\right)}2E\left(X^3\right) + 3E\left(X^2\right)$$

$$9[E\left(X^2\right)]^2 + 24\theta\mu E\left(X^3\right) \quad < \quad \left[\frac{4\theta\mu}{E\left(X^2\right)}E\left(X^3\right) + 3E\left(X^2\right)\right]^2$$

$$9[E\left(X^2\right)]^2 + 24\theta\mu E\left(X^3\right) \quad < \quad \left[\frac{4\theta\mu}{E\left(X^2\right)}E\left(X^3\right)\right]^2$$

$$+ 24\theta\mu E\left(X^3\right) + 9[E\left(X^2\right)]^2$$

which implies $0 < \left[\frac{4\theta\mu}{E(X^2)}E\left(X^3\right)\right]^2$

6.11 (a) $1 + \theta = \int_0^\infty e^{\kappa x} f_e(x)dx$ where $f_e(x) = \frac{1-f(x)}{\mu}, x > 0$.

Let $g(x) = e^x$ then $g''(x) = e^x \geq 0$ and then by Jensen's inequality,

$$1 + \theta \quad = \quad \int_0^\infty e^{\kappa x} f_e(x)dx = E\left(e^{\kappa X}\right)$$

$$\geq \quad e^{E(\kappa X)} = e^{\int_0^\infty \kappa x f_e(x)dx}$$
$$= \quad e^{\kappa E(X^2)/(2\mu)}.$$

Therefore, $\log(1+\theta) \geq \kappa E(X^2)/(2\mu)$ and so $\kappa \leq [2\mu \log(1+\theta)]/E(X^2)$.

(b) Consider $h(\theta) = \theta - \log(1+\theta)$, $\theta \geq 0$. Then $h'(\theta) = 1 - \frac{1}{1+\theta} > 0$ for $\theta > 0$. Thus $h(\theta)$ is absolutely increasing in $(0, \infty)$.

Since $h(0) = 0$, $h(\theta) > h(0) = 0$, $\theta > 0$, that is, $\theta > \log(1+\theta)$ for $\theta > 0$.

Hence, $\leq \frac{2\mu \log(1+\theta)}{E(X^2)} < \frac{2\mu\theta}{E(X^2)}$.

(c) If there is a maximum claim size of m, then $X \leq m$ and (6.8) becomes

$$1+\theta \quad = \quad \int_0^\infty e^{\kappa x} f_e(x)dx = \int_0^m e^{\kappa x} f_e(x)dx$$
$$\leq \quad \int_0^m e^{\kappa m} f_e(x)dx = e^{\kappa m} \int_0^m f_e(x)dx = e^{\kappa m}.$$

So $1+\theta \leq e^{\kappa m}$ and therefore, $\kappa \geq \frac{1}{m}\log(1+\theta)$.

6.12 (a) We have $S_e(x) \geq S(x)$, or since the mean residual lifetime $e(x)$ satisfies

$$\frac{e(x)}{e(0)} = \frac{S_e(x)}{S(x)},$$

equivalently, $e(x) \geq e(0)$. Now, $S_e(y) \geq S(y)$ may be stated as $\Pr(Y > y) \geq \Pr(X > y)$. Since the function $e^{\kappa x}$ is monotone increasing in x, this is equivalent to stating that $\Pr\left(e^{\kappa Y} > e^{\kappa y}\right) \geq \Pr\left(e^{\kappa X} > e^{\kappa y}\right)$, or $\Pr\left(e^{\kappa Y} > t\right) \geq \Pr\left(e^{\kappa X} > t\right)$ where $t = e^{\kappa y} \geq 1$.

(b) For any non-negative random variable W, $E(W) = \int_0^\infty \Pr(W > t)dt$. Thus, with $W = e^{\kappa Y}$ we have from (a)

$$E\left(e^{\kappa Y}\right) = \int_0^\infty \Pr\left(e^{\kappa Y} > t\right) dt \geq \int_0^\infty \Pr\left(e^{\kappa X} > t\right) dt = E\left(e^{\kappa X}\right).$$

(c) From (b) and (6.8),

$$1+\theta = E\left(e^{\kappa Y}\right) \geq E\left(e^{\kappa X}\right) = 1 + (1+\theta)\mu\kappa.$$

Thus,

$$\kappa \leq \frac{\theta}{\mu(1+\theta)}.$$

(d) Simply reverse the inequalities in (a), (b), and (c).

6.13 Let $\psi_n(u)$ be the probability that ruin occurs on or before the nth claim for $n = 0, 1, 2, \cdots$. We will prove by induction on n that $\psi_n(u) \leq \rho e^{-\kappa u}$.

Obviously $\psi_0(u) = 0 \le \rho e^{-\kappa u}$. Now assume that $\psi_n(u) \le \rho e^{-\kappa u}$. Then

$$
\begin{aligned}
\psi_{n+1}(u) &= \int_0^\infty \left[1 - F(u+ct) + \int_0^{u+ct} \psi_n(u+ct-x)dF(x) \right] \lambda e^{-\lambda t}dt \\
&= \int_0^\infty \left[\overline{F}(u+ct) + \int_0^{u+ct} \psi_n(u+ct-x)dF(x) \right] \lambda e^{-\lambda t}dt \\
&\le \int_0^\infty \left[\rho e^{-\kappa(u+ct)} \int_{u+ct}^\infty e^{\kappa x}dF(x) \right. \\
&\qquad \left. + \int_0^{u+ct} \rho e^{-\kappa(u+ct-X)}dF(x) \right] \lambda e^{-\lambda t}dt \\
&= \rho \int_0^\infty \left[\int_0^\infty e^{-\kappa(u+ct-X)}dF(x) \right] \lambda e^{-\lambda t}dt \\
&= \rho e^{-\kappa u} \int_0^\infty e^{-\kappa ct} \left[\int_0^\infty e^{\kappa X}dF(x) \right] \lambda e^{-\lambda t}dt \\
&= \rho \lambda e^{-\kappa u} \int_0^\infty e^{-\lambda t - \kappa ct} E\left(e^{\kappa X} \right) dt \\
&= \rho \lambda e^{-\kappa u} E\left(e^{\kappa X} \right) \int_0^\infty e^{-(\lambda + \kappa c)t}dt \\
&= \rho \lambda e^{-\kappa u} E\left(e^{\kappa X} \right) / (\lambda + \kappa c) \\
&= \rho e^{-\kappa u}
\end{aligned}
$$

since $\lambda E\left(e^{\kappa X} \right) = \lambda[1+(1+\theta)\kappa\mu] = \lambda + \kappa(1+\theta)\lambda\mu = \lambda + \kappa c$. So $\psi_n(u) \le \rho e^{-\kappa u}$, $n = 0, 1, 2, \cdots$, which implies $\psi(u) = \lim_{n\to\infty} \psi_n(u) \le \rho e^{-\kappa u}$

6.14
$$
\begin{aligned}
\int_x^\infty e^{\kappa y}dF(y) &= -\int_x^\infty e^{\kappa y}d\overline{F}(y) \\
&= -e^{\kappa y}\overline{F}(y)\Big|_x^\infty + \kappa \int_x^\infty e^{\kappa y}\overline{F}(y)dy \quad X \ge 0 \\
&= -\lim_{y\to\infty} e^{\kappa y}\overline{F}(y) + e^{\kappa x}\overline{F}(x) + \kappa \int_x^\infty e^{\kappa y}\overline{F}(y)dy
\end{aligned}
$$

Since

$$
\begin{aligned}
0 &\le e^{\kappa y}\overline{F}(y) = e^{\kappa y}[1 - F(y)] \\
&= e^{\kappa y}\int_y^\infty dF(x) \le \int_y^\infty e^{\kappa x}dF(x) \\
&\le \int_0^\infty e^{\kappa x}dF(x) = E\left(e^{\kappa X} \right) < \infty
\end{aligned}
$$

and $\lim_{y\to\infty} e^{\kappa y}\overline{F}(y) \le \lim_{y\to\infty} \int_y^\infty e^{\kappa x}dF(x) = 0$, we have $\lim_{y\to\infty} e^{\kappa y}\overline{F}(y) = 0$.

6.15 Let $\mu(x) = \frac{f(x)}{\overline{F}(x)}$ be the failure rate. Then $\mu(x+t) \le \mu(x)$ for each x and $t \ge 0$ by assumption, so $\frac{f(x+t)}{\overline{F}(x+t)} \le \frac{f(x)}{\overline{F}(x)}$ and therefore $f(x+t)\overline{F}(x) - f(x)\overline{F}(x+t) < 0$ for all x, and all $t \ge 0$. For each fixed $t \ge 0$, consider $g_t(x) = \overline{F}(x+t)/\overline{F}(x)$. Then

$$\frac{d}{dx}g_t(x) = \frac{-\overline{F}(x)f(x+t) + \overline{F}(x+t)f(x)}{[\overline{F}(x)]^2} \ge 0$$

which implies that $g_t(x)$ is increasing in x for each fixed t. So $g_t(x) \ge g_t(0)$, and therefore

$$\frac{\overline{F}(x+t)}{\overline{F}(x)} \ge \frac{\overline{F}(t)}{\overline{F}(0)} = \overline{F}(t), \text{ or } \overline{F}(x+t) \ge \overline{F}(x)\overline{F}(t).$$

Hence $\overline{F}(y) = \overline{F}[x + (y - x)] \ge \overline{F}(x)\overline{F}(y - x)$, for all $x \ge 0$ and $y \ge x$. then, by Exercise 6.14,

$$
\begin{aligned}
\rho e^{-\kappa x}\int_x^\infty e^{\kappa y}dF(y) &= \rho e^{-\kappa x}\left[e^{\kappa x}\overline{F}(x) + \kappa \int_x^\infty e^{\kappa y}\overline{F}(y)dy\right] \\
&\ge \rho\overline{F}(x) + \rho\kappa e^{-\kappa x}\int_x^\infty e^{\kappa y}\overline{F}(x)\overline{F}(y - x)dy \\
&= \rho\overline{F}(x) + \rho\overline{F}(x)\kappa \int_x^\infty e^{\kappa(y-x)}\overline{F}(y - x)dy \\
&= \rho\overline{F}(x)\left[1 + \kappa \int_0^\infty e^{\kappa z}\overline{F}(z)dz\right], \text{ where } z = y - x, \\
&= \rho\overline{F}(x)\left[1 + \int_0^\infty \overline{F}(x)de^{\kappa x}\right] \\
&= \rho\overline{F}(x)\left[1 + \overline{F}(x)e^{\kappa x}\Big|_0^\infty + \int_0^\infty e^{\kappa x}f(x)dx\right] \\
&= \rho\overline{F}(x)\left[1 + \lim_{x\to\infty}\overline{F}(x)e^{\kappa x} - 1 + E\left(e^{\kappa X}\right)\right] \\
&= \rho\overline{F}(x)E\left(e^{\kappa X}\right)
\end{aligned}
$$

If $\rho^{-1} = E\left(e^{\kappa X}\right)$, then $\rho e^{-\kappa x}\int_x^\infty e^{\kappa y}dF(y) \ge \rho\overline{F}(x)E\left(e^{\kappa X}\right) = \overline{F}(x)$. (6.65) is satisfied. From Exercise 6.13,

$$\psi(x) \le \rho e^{-\kappa x} = \left[E\left(e^{\kappa X}\right)\right]^{-1}e^{-\kappa x}$$

6.16 $\mu(x) = -\frac{d}{dx}\log\overline{F}(x) \Rightarrow \overline{F}(x) = e^{-\int_0^x \mu(t)dt}$. For $y > x$,

$$
\begin{aligned}
\frac{\overline{F}(y)}{\overline{F}(x)} &= \frac{e^{-\int_0^y \mu(t)dt}}{e^{-\int_0^x \mu(t)dt}} = e^{-\int_x^y \mu(t)dt} \\
&\ge e^{-\int_x^y mdt} = e^{-(y-x)m}
\end{aligned}
$$

which implies $\overline{F}(y) \geq \overline{F}(x)e^{-(y-x)m}$. From Exercise 6.14,

$$
\begin{aligned}
\rho e^{-\kappa x} \int_x^\infty e^{\kappa y} dF(y) &= \rho e^{-\kappa x} \left[e^{\kappa x} \overline{F}(x) + \kappa \int_x^\infty e^{\kappa y} \overline{F}(y) dy \right] \\
&\geq \rho \overline{F}(x) \left[+ \rho \kappa e^{-\kappa X} \int_x^\infty e^{\kappa y} \overline{F}(x) e^{-(y-x)m} dy \right] \\
&= \rho \overline{F}(x) \left[1 + \kappa \int_x^\infty e^{\kappa(y-x)} e^{-(y-x)m} dy \right] \\
&= \rho \overline{F}(x) \left[1 + \kappa \int_0^\infty e^{-(m-\kappa)z} dz \right], \quad \text{where } z = y - x, \\
&= \rho \overline{F}(x) \left[1 - \frac{\kappa}{m-\kappa} e^{-(m-\kappa)z} \Big|_0^\infty \right] \\
&= \rho \overline{F}(x) \left[1 + \frac{\kappa}{m-\kappa} \right], \quad \text{if } m > \kappa, \\
&= \rho \overline{F}(x) \frac{m}{m-\kappa}.
\end{aligned}
$$

If $\rho = \left(\frac{m}{m-\kappa} \right)^{-1} = 1 - \frac{\kappa}{m}$, then $\rho e^{-\kappa x} \int_x^\infty e^{\kappa y} dF(y) \geq \rho \overline{F}(x) \frac{m}{m-\kappa} = \overline{F}(x)$, (6.65) is satisfied. From Exercise 6.13, $\psi(x) \leq \rho e^{-\kappa x} = \left(1 - \frac{\kappa}{m}\right) e^{-\kappa x}$, $x \geq 0$.

5.4   SECTION 6.6

6.17
$$
\begin{aligned}
\mu &= E(X) = \int_0^\infty x e^{-3x} dx + 6 \int_0^\infty x^2 e^{-3x} dx \\
&= \Gamma(12) \left(\frac{1}{3} \right)^2 + 6\Gamma(3) \left(\frac{1}{3} \right)^3 = \frac{5}{9}
\end{aligned}
$$

$\frac{c}{\lambda} = \mu(1+\theta) = \frac{5}{9}\left(1 + \frac{4}{5}\right) = 1$. Equation (6.18) becomes

$$
\begin{aligned}
\phi'(u) &= \phi(u) - \int_0^u \phi(u-x)(1+6x)e^{-3x} dx \\
&= \phi(u) - \int_0^u \phi(y) \left[1 + 6(u-y)\right] e^{-3(u-y)} dy, \quad \text{where } y = u - x \\
&= \phi(u) - (1+6u)e^{-3u} r_3(u) + 6e^{-3u} \int_0^u y\phi(y)e^{3y} dy \qquad (5.1)
\end{aligned}
$$

where $r_n(u) = \int_0^u e^{ny} \phi(y) dy$. Then

$$
\begin{aligned}
\phi''(u) &= \phi'(u) + 3(1+6u)e^{-3u} r_3(u) - 6e^{-3u} r_3(u) \\
&\quad - (1+6u)e^{-3u} e^{3u} \phi(u) - 18e^{-3u} \int_0^u y\phi(y)e^{3y} dy
\end{aligned}
$$

$$+6e^{-3u}u\phi(u)e^{3u}$$

$$= \phi'(u) + 3(6u-1)e^{-3u}r_3(u) - \phi(u) - 18e^{-3u}\int_0^u y\phi(y)e^{3y}dy$$

$$= \phi'(u) + 3(6u-1)e^{-3u}r_3(u) - \phi(u)$$
$$-3\left[\phi'(u) - \phi(u) + (1+6u)e^{-3u}r_3(u)\right] \text{ from (5.1)}$$

$$= 2\left[\phi(u) - \phi'(u)\right] - 6e^{-3u}r_3(u) \qquad (5.2)$$

$$\phi''(u) = 2\left(\phi'(u) - \phi''(u)\right) + 18e^{-3u}r_3(u) - 6e^{-3u}e^{3u}\phi(u)$$

$$= 2\left(\phi'(u) - \phi''(u)\right) + 3\left[2\left(\phi(u) - \phi'(u)\right) - \phi''(u)\right]$$
$$-6\phi(u) \text{ from (5.2)}$$

$$= -5\phi''(u) - 4\phi'(u)$$

So $\phi''(u) + 5\phi''(u) + 4\phi'(u) = 0$. Since $0 = r^2 + 5r + 4 = (r+1)(r+4)$ has roots $r_1 = -1$ and $r_2 = -4$, we have $\phi'(u) = A_1e^{-u} + A_2e^{-4u}$. Then

$$A_1 + A_2 = \phi'(0) = \phi(0) = \frac{\theta}{1+\theta} = \frac{4}{9}. \qquad (5.3)$$

Also, $\phi(u) = -A_1e^{-u} - \frac{1}{4}A_2e^{-4u} + A_3$. But then, $\phi(\infty) = 1 \Rightarrow A_3 = 1$ and so $1 - A_1 - \frac{1}{4}A_2 = \phi(0) = \frac{4}{9}$ which implies

$$A_1 + \frac{1}{4}A_2 = \frac{5}{9}. \qquad (5.4)$$

Then (5.3) and (5.4) imply $A_1 = \frac{16}{27}$ and $A_2 = -\frac{4}{27}$. Hence $\phi(u) = 1 - \frac{16}{27}e^{-u} + \frac{1}{27}e^{-4u}$.

6.18
$$\mu = E(X) = \frac{1}{2}\left[\int_0^\infty 4xe^{-4x}dx + \int_0^\infty 7xe^{-7x}dx\right]$$

$$= \frac{1}{2}\left(\frac{1}{4} + \frac{1}{7}\right) = \frac{11}{56}.$$

Also,

$$\frac{c}{\lambda} = \mu(1+\theta) = \frac{11}{56}\frac{14}{11} = \frac{1}{4}.$$

Equation (6.18) becomes

$$\phi'(u) = 4\phi(u) - 4\int_0^u \phi(u-x)\left[2e^{-4x} + \frac{7}{2}e^{-7x}\right]dx$$

$$= 4\phi(u) - 4\int_0^u \phi(y)\left[2e^{-4(u-y)} + \frac{7}{2}e^{-7(u-y)}\right]dy, \text{ where } y = u - x$$

$$= 4\phi(u) - 8e^{-4u}r_4(u) - 14e^{-7u}r_7(u) \qquad (5.5)$$

where $r_n(u) = \int_0^u e^{ny}\phi(y)dy$. Then,

$$
\begin{aligned}
\phi''(u) &= 4\phi'(u) + 32e^{-4u}r_4(u) - 8\phi(u) \\
&\quad +98e^{-7u}r_7(u) - 14\phi(u) \\
&= 4\phi'(u) - 22\phi(u) + 32e^{-4u}r_4(u) + 98e^{-7u}r_7(u) \\
&= 4\phi'(u) - 22\phi(u) + 32e^{-4u}r_4(u) \\
&\quad +7\left[4\phi(u) - 8e^{-4u}r_4(u) - \phi'(u)\right] \text{ from (5.5)} \\
&= -3\phi'(u) + 6\phi(u) - 24e^{-4u}r_4(u) \qquad\qquad (5.6)
\end{aligned}
$$

and

$$
\begin{aligned}
\phi'''(u) &= -3\phi''(u) + 6\phi'(u) + 96e^{-4u}r_4(u) - 24\phi(u) \\
&= -3\phi''(u) + 6\phi'(u) \\
&\quad +4\left[-3\phi'(u) + 6\phi(u) - \phi''(u)\right] \\
&\quad -24\phi(u) \text{ from (5.6)} \\
&= -7\phi''(u) - 6\phi'(u)
\end{aligned}
$$

So

$$
\phi'''(u) + 7\phi''(u) + 6\phi'(u) = 0.
$$

Since $0 = r^2 + 7r + 6 = (r+6)(r+1)$ has roots $r_1 = -6$ and $r_2 = -1$, we have $\phi'(u) = A_1 e^{-6u} + A_2 e^{-u}$. Then,

$$
A_1 + A_2 = \phi'(0) = 4\phi(0) = 4\frac{\theta}{1+\theta} = \frac{6}{7}. \qquad (5.7)
$$

Also, $\phi(u) = -\frac{1}{6}A_1 e^{-6u} - A_2 e^{-u} + A_3$. and since $\phi(\infty) = 1$ we have $A_3 = 1$. Then

$$
1 - \frac{1}{6}A_1 - A_2 = \phi(0) = \frac{3}{14} \Rightarrow \frac{1}{6}A_1 + A_2 = \frac{11}{14}. \qquad (5.8)
$$

From (5.7) and (5.8) we have $A_1 = \frac{3}{35}, A_2 = \frac{27}{35}$. Hence, $\phi(u) = 1 - \frac{1}{70}e^{-6u} - \frac{27}{35}e^{-u}$.

6.19
$$
\begin{aligned}
\mu &= E(X) = \frac{3}{4}\int_0^\infty 4xe^{-4x}dx + \frac{1}{4}\int_0^\infty 2xe^{-2x}dx \\
&= \frac{3}{4}\frac{1}{4} + \frac{1}{4}\frac{1}{2} = \frac{5}{16}.
\end{aligned}
$$

$$
\frac{c}{\lambda} = \mu(1+\theta) = \frac{5}{16} \times \frac{8}{5} = \frac{1}{2}.
$$

Equation (6.18) becomes

$$
\begin{aligned}
\phi'(u) &= 2\phi(u) - 2\int_0^u \phi(u-x)\left(3e^{-4x} + \frac{1}{2}e^{-2x}\right)dx \\
&= 2\phi(u) - 2\int_0^u \phi(y)\left[3e^{-4(u-y)} + \frac{1}{2}e^{-2(u-y)}\right]dy, \text{ where } y = u - x \\
&= 2\phi(u) - 6e^{-4u}r_4(u) - e^{-2u}r_2(u) \qquad\qquad (5.9)
\end{aligned}
$$

where $r_n(u) = \int_0^u e^{ny}\phi(y)dy$.

$$
\begin{aligned}
\phi''(u) &= 2\phi'(u) + 24e^{-4u}r_4(u) - 6\phi(u) \\
&\quad + 2e^{-2u}r_2(u) - \phi(u) \\
&= 2\phi'(u) - 7\phi(u) + 24e^{-4u}r_4(u) \\
&\quad + 2\left[2\phi(u) - 6e^{-4u}r_4(u) - \phi'(u)\right] \text{ from } (5.9) \\
&= -3\phi(u) + 12e^{-4u}r_4(u). \tag{5.10}
\end{aligned}
$$

$$
\begin{aligned}
\phi'''(u) &= -3\phi'(u) - 48e^{-4u}r_4(u) + 12\phi(u) \\
&= -3\phi'(u) + 12\phi(u) \\
&\quad -4\left[\phi''(u) + 3\phi(u)\right] \text{ from } (5.10) \\
&= -4\phi''(u) - 3\phi'(u)
\end{aligned}
$$

So

$$
\phi'''(u) + 4\phi''(u) + 3\phi'(u) = 0.
$$

Since $0 = r^2 + 4r + 3 = (r+1)(r+3)$ has roots $r_1 = -1$ and $r_2 = -3$, we have $\phi'(u) = A_1 e^{-u} + A_2 e^{-3u}$. Then

$$
A_1 + A_2 = \phi'(0) = 2\phi(0) = 2\frac{\theta}{1+\theta} = \frac{3}{4} \tag{5.11}
$$

Also, $\phi(u) = -A_1 e^{-u} - \frac{1}{3}A_2 e^{-3u} + A_3$. Becaue $\phi(\infty) = 1$ we have $A_3 = 1$. Then

$$
1 - A_1 - \frac{1}{3}A_2 = \phi(0) = \frac{\theta}{1+\theta} = \frac{3}{8} \Rightarrow A_1 + \frac{1}{3}A_2 = \frac{5}{8} \tag{5.12}
$$

Then (5.11) and (5.12) imply $A_1 = \frac{9}{16}$, $A_2 = \frac{3}{16}$.   Hence, $\phi(u) = 1 - \frac{9}{16}e^{-u} - \frac{1}{16}e^{-3u}$.

6.20

$$
\begin{aligned}
f(x) &= \int_0^x 3e^{-3x}2e^y dy = 6e^{-3x}\int_0^x e^y dy \\
&= 6e^{-3x}(e^x - 1) = 6e^{-2x} - 6e^{-3x} \\
\mu &= E(X) = 3\int_0^\infty 2xe^{-2x}dx - 2\int_0^\infty 3xe^{-3x}dx \\
&= \frac{3}{2} - \frac{2}{3} = \frac{5}{6} \\
\frac{c}{\lambda} &= \mu(1+\theta) = \frac{5}{6} \times \frac{12}{5} = 2
\end{aligned}
$$

Equation (6.18) becomes

$$
2\phi'(u) = \phi(u) - \int_0^u \phi(u-x)(6e^{-2x} - 6e^{-3x})dx \tag{5.13}
$$

$$= \quad \phi(u) - 6 \int_0^u \phi(y) \left[e^{-2(u-y)} - e^{-3(u-y)} \right] dy, \text{ where } y = u - x$$

$$\phi(u) - 6e^{-2u} r_2(u) + 6e^{-3u} r_3(u)$$

where $r_n(u) = \int_0^u e^{ny} \phi(y) dy$.

$$
\begin{aligned}
2\phi''(u) &= \phi'(u) + 12e^{-2u} r_2(u) - 6\phi(u) \\
&\quad -18e^{-3u} r_3(u) + 6\phi(u) \\
&= \phi'(u) + 12e^{-2u} r_2(u) \\
&\quad -3 \left[2\phi'(u) - \phi(u) + 6e^{-2u} r_2(u) \right] \text{ from (5.13)} \\
&= -5\phi'(u) + 3(u) - 6e^{-2u} r_2(u). \quad (5.14)
\end{aligned}
$$

$$
\begin{aligned}
2\phi'''(u) &= -5\phi''(u) + 3\phi'(u) \\
&\quad +12e^{-2u} r_2(u) - 6\phi(u) \\
&= 5\phi''(u) + 3\phi'(u) \\
&\quad +2\left[-5\phi'(u) + 3\phi(u) - 2\phi''(u) \right] - 6\phi(u) \\
&= -9\phi''(u) - 7\phi'(u) \text{ from (5.14)}.
\end{aligned}
$$

So
$$2\phi'''(u) + 9\phi'''(u) + 7\phi'(u) = 0.$$

Also, $0 = 2r^2 + 9r + 7 = (2r + 7)(r + 1) = 0$ has roots $r_1 = -\frac{7}{2}$ and $r_2 = -1$, and so $\phi'(u) = A_1 e^{-\frac{7}{2}u} + A_2 e^{-u}$. Then

$$A_1 + A_2 = \phi'(0) = \frac{1}{2}\phi(0) = \frac{1}{2}\frac{\theta}{1+\theta} = \frac{7}{24}. \quad (5.15)$$

$\phi(u) = -\frac{2}{7}A_1 e^{-\frac{7}{2}u} - A_2 e^{-u} + A_3$ and $\phi(\infty) = 1$ implies $A_3 = 1$. Then,

$$1 - \frac{2}{n}A_1 - A_2 = \frac{5}{12} \quad (5.16)$$

Then (5.15) and (5.16) imply $A_1 = -\frac{21}{120}$, $A_2 = \frac{56}{120} = \frac{7}{15}$. Hence, $\phi(u) = 1 + \frac{1}{20}e^{-\frac{7}{2}u} - \frac{7}{15}e^{-u}$.

6.21 (a)

$$
\begin{aligned}
c\phi'(u) &= \lambda\phi(u) - \lambda \int_0^u \phi(u - x) dF(x) \quad u \geq 0 \\
&= \lambda\phi(u) + \lambda \int_0^u \phi(u - x) d\overline{F}(x) \\
&= \lambda\phi(u) + \lambda \left[\phi(u - x)\overline{F}(x)\Big|_0^u + \int_0^u \phi'(u - x)\overline{F}(x) dx \right]
\end{aligned}
$$

$$= \lambda\phi(u) + \lambda\left[\phi(0)\overline{F}(u) - \phi(u)\right] + \lambda\int_0^u \phi'(u-x)\overline{F}(x)dx$$

$$= \lambda\phi(0)\left[1 - F(u)\right] + \lambda\int_0^u \phi'(u-x)\left[1 - F(x)\right]dx.$$

(b) $ce^{-\kappa u}m(u) = \lambda\phi(0)\left[1 - F(u)\right] + \lambda\int_0^u \phi'(u-x)\left[1 - F(x)\right]dx$

$$
\begin{aligned}
m(u) &= \frac{\lambda}{c}e^{\kappa u}\phi(0)\left[1 - F(u)\right] \\
&\quad + \frac{\lambda}{c}e^{\kappa u}\int_0^u e^{-\kappa(u-x)}m(u-x)\left[1 - F(x)\right]dx \\
&= \frac{1}{\mu(1+\theta)}e^{\kappa u}\frac{\theta}{1+\theta}\left[1 - F(u)\right] \\
&\quad + \frac{1}{\mu(1+\theta)}\int_0^u m(u-x)e^{\kappa x}\left[1 - F(x)\right]dx \\
&= \frac{\theta}{(1+\theta)^2}e^{\kappa u}f_e(u) + \frac{1}{1+\theta}\int_0^u m(u-x)e^{\kappa x}f_e(x)dx,
\end{aligned}
$$

where $f_e(u) = \dfrac{1 - F(u)}{\mu}$.

(c) $f(x) = \frac{1}{3}\left[2e^{-2x} + 3e^{-3x} + 6e^{-6x}\right]$ $x \geq 0$

$$
\begin{aligned}
\mu &= E(x) = \frac{1}{3}\left[\int_0^\infty 2xe^{-2x}dx + \int_0^\infty 3xe^{-3x}dx + \int_0^\infty 6xe^{-6x}dx\right] \\
&= \frac{1}{3}\left(\frac{1}{2} + \frac{1}{3} + \frac{1}{6}\right) = \frac{1}{3}.
\end{aligned}
$$

$1.7 = c = (1+\theta)\lambda\mu = (1+\theta)3(1/3)$ which implies $\theta = 0.7$ From (6.4),

$$
\begin{aligned}
1 + \frac{1.7}{3}\kappa &= E\left(e^{\kappa X}\right) = \frac{1}{3}\left[2\int_0^\infty e^{-(2-\kappa)x}dx\right. \\
&\quad \left. + 3\int_0^\infty e^{-(3-\kappa)x}dx + 6\int_0^\infty e^{-(6-\kappa)x}dx\right] \\
&= \frac{1}{3}\left(\frac{2}{2-\kappa} + \frac{3}{3-\kappa} + \frac{6}{6-\kappa}\right).
\end{aligned}
$$

Rewrite this equation as

$$
\begin{aligned}
3 + 1.7\kappa &= \frac{2}{2-\kappa} + \frac{3}{3-\kappa} + \frac{6}{6-\kappa} \\
&= \frac{11\kappa^2 - 72\kappa + 108}{(2-\kappa)(3-\kappa)(6-\kappa)}
\end{aligned}
$$

which implies

$$
\begin{aligned}
0 &= 17\kappa^4 - 157\kappa^3 + 392\kappa^2 - 252\kappa \\
&= \kappa(\kappa-1)\left(17\kappa^2 - 140\kappa + 252\right)
\end{aligned}
$$

and the roots are $0, 1, \frac{70 \pm 2\sqrt{154}}{17} = 2.6577$ and 5.5776. The adjustment coefficient is $\kappa = 1$, the smallest positive root. We next have

$$
\begin{aligned}
F(x) &= \frac{1}{3}\left(1 - e^{-2x} + 1 - e^{-3x} + 1 - e^{-6x}\right) \\
&= 1 - \frac{1}{3}\left(e^{-2x} + e^{-3x} + e^{-6x}\right) \\
f_e(x) &= \frac{1 - F(x)}{\mu} = e^{-2x} + e^{-3x} + e^{-6x} \\
m(u) &= \frac{\theta}{(1+\theta)^2} e^{\kappa u}\left(e^{-2u} + e^{-3u} + e^{-6u}\right) \\
&\quad + \frac{1}{1+\theta}\int_0^u m(u-x)e^{\kappa x}\left(e^{-2x} + e^{-3x} + e^{-6x}\right)dx \\
&= \frac{0.7}{(1.7)^2}\left(e^{-u} + e^{-2u} + e^{-5u}\right) \\
&\quad + \frac{1}{1.7}\int_0^u m(u-x)\left(e^{-x} + e^{-2x} + e^{-5x}\right)dx \\
&= \frac{0.7}{(1.7)^2}\left(e^{-u} + e^{-2u} + e^{-5u}\right) \\
&\quad + \frac{1}{1.7}\int_0^u m(y)\left[e^{y-u} + e^{2(y-u)} + e^{5(y-u)}\right]dy, \text{ where } y = u - x \\
&= \frac{0.7}{2.89}\left(e^{-u} + e^{-2u} + e^{-5u}\right) \\
&\quad + \frac{1}{1.7}\left[e^{-u}s_1(u) + e^{-2u}s_2(u) + e^{-5u}s_5(u)\right] \qquad (5.17)
\end{aligned}
$$

where $s_n(u) = \int_0^u m(y)e^{ny}dy$.

$$
\begin{aligned}
m'(u) &= -\frac{0.7}{2.89}\left(e^{-u} + 2e^{-2u} + 5e^{-5u}\right) \\
&\quad + \frac{1}{1.7}\left[-e^{-u}s_1(u) + m(u) - 2e^{-2u}s_2(u)\right. \\
&\quad \left. + m(u) - 5e^{-5u}s_5(u) + m(u)\right] \\
&= -\frac{0.7}{2.89}\left(e^{-u} + 2e^{-2u} + 5e^{-5u}\right) \\
&\quad + \frac{3}{1.7}m(u) - \frac{1}{1.7}\left[e^{-u}s_1(u) + 2e^{-2u}s_2(u) + 5e^{-5u}s_5(u)\right] \\
&= -\frac{0.7}{2.89}\left(e^{-u} + 2e^{-2u} + 5e^{-5u}\right) + \frac{3m(u)}{1.7} \\
&\quad - \frac{1}{1.7}\left[e^{-u}s_1(u) + 2e^{-2u}s_2(u)\right] \\
&\quad - 5\left\{m(u) - \frac{0.7}{2.89}\left(e^{-u} + e^{-2u} + e^{-5u}\right)\right.
\end{aligned}
$$

$$-\frac{1}{1.7}\left[e^{-u}s_1(u)+e^{-2u}s_2(u)\right]\Big\}\ \text{from (5.17)}$$

$$=\ \frac{1}{2.89}\left(2.8e^{-u}+2.1e^{-2u}\right)-\frac{5.5}{1.7}m(u)$$

$$+\frac{4}{1.7}e^{-u}s_1(u)+\frac{3}{1.7}e^{-2u}s_2(u) \tag{5.18}$$

$$m''(u)\ =\ -\frac{1}{2.89}\left(2.8e^{-u}+4.2e^{-2u}\right)-\frac{5.5}{1.7}m'(u)$$

$$-\frac{4}{1.7}e^{-u}s_1(u)+\frac{4}{1.7}m(u)-\frac{6}{1.7}e^{-2u}s_2(u)$$

$$+\frac{3}{1.7}m(u)$$

$$=\ \frac{1}{2.89}\left(2.8e^{-u}+4.2e^{-2u}\right)-\frac{5.5}{1.7}m'(u)$$

$$+\frac{7}{1.7}m(u)-\frac{4}{1.7}e^{-u}s_1(u)-2\left[m'(u)-\frac{1}{2.89}\left(2.8e^{-u}+2.1e^{-2u}\right)\right.$$

$$\left.+\frac{5.5}{1.7}m(u)-\frac{4}{1.7}e^{-u}s_1(u)\right]\ \text{from (5.18)}$$

$$=\ \frac{2.8}{2.89}e^{-u}-\frac{8.9}{1.7}m'(u)-\frac{4}{1.7}m(u)+\frac{4}{1.7}e^{-u}s_1(u) \tag{5.19}$$

$$m'''(u)\ =\ -\frac{2.8}{2.89}e^{-u}-\frac{8.9}{1.7}m''(u)-\frac{4}{1.7}m'(u)-\frac{4}{1.7}e^{-u}s_1(u)+\frac{4}{1.7}m(u)$$

$$=\ -\frac{2.8}{2.89}e^{-u}-\frac{8.9}{1.7}m''(u)-\frac{4}{1.7}m'(u)$$

$$+\left[\frac{2.8}{2.89}e^{-u}-\frac{8.9}{1.7}m'(u)-\frac{4}{1.7}m(u)-m''(u)\right]+\frac{4}{1.7}m(u)$$

$$=\ -\frac{10.6}{1.7}m''(u)-\frac{12.9}{1.7}m'(u)\ \text{from (5.19).}$$

So

$$17m'''(u)+106m''(u)+129m'(u)=0.$$

Since $17r^2+106r+129=0$ has roots $r_1=\frac{-53-2\sqrt{154}}{17}$ and $r_2\frac{-53+2\sqrt{154}}{17}$, we have $m'(u)=A_1e^{r_1u}+A_2^{r_2u}$. Also,

$$A_1+A_2\ =\ m'(0)=-\frac{0.7}{2.89}(1+2+5)+\frac{3}{1.7}m(0)$$

$$=\ -\frac{5.6}{2.89}+\frac{3}{1.7}\times\frac{0.7}{2.89}\times(1+1+1)$$

$$=\ -\frac{3.22}{(1.7)^3}=-\frac{3220}{4913}. \tag{5.20}$$

$m(u) = \frac{A_1}{r_1} e^{r_1 u} + \frac{A_2}{r_2} e^{r_2 u} + A_3$ which implies $A_3 = \lim_{u \to \infty} m(u) = c\kappa = \frac{\theta \mu \kappa}{E(Xe^{\kappa X}) - \mu(1+\theta)}$ by Exercise 6.25. Then

$$E\left(Xe^{\kappa X}\right) = \frac{d}{d\kappa} E\left(e^{\kappa X}\right) = \frac{d}{d\kappa}\left[\frac{1}{3}\left(\frac{2}{2-\kappa} + \frac{3}{3-\kappa} + \frac{6}{6-\kappa}\right)\right]$$

$$= \frac{1}{3}\left[\frac{2}{(2-\kappa)^2} + \frac{3}{(3-\kappa)^2} + \frac{6}{(6-\kappa)^2}\right]$$

and so $E\left(Xe^{\kappa X}\right)\big|_{\kappa=1} = \frac{1}{3}\left(2 + \frac{3}{4} + \frac{6}{25}\right) = \frac{299}{300}$. We then have

$$A_3 = c\kappa = \frac{0.7 \times \frac{1}{3} \times 1}{299/300 - 1.7/3} = \frac{70}{129}$$

and so

$$A_3 + \frac{A_1}{r_1} + \frac{A_2}{r_2} = m(0) = \frac{0.7}{2.89} \times 3 = \frac{2.1}{289} = \frac{210}{289}$$

which implies

$$\frac{A_1}{r_1} + \frac{A_2}{r_2} = \frac{210}{289} - A_3 = \frac{6860}{37281} \tag{5.21}$$

Then (5.20) and (5.21) imply that

$$A_1 = -\frac{1610}{4913} - \frac{965\sqrt{154}}{54043} \text{ and } A_2 = -\frac{1610}{4913} + \frac{965\sqrt{154}}{54043}.$$

Then $m(u) = \frac{70}{129} + \frac{A_1}{r_1} e^{r_1 u} + \frac{A_2}{r_2} e^{r_2 u}$ where $r_1 = \frac{-53-2\sqrt{154}}{17}$ and $r_2 = \frac{-53+2\sqrt{154}}{17}$.

5.5   SECTION 6.7

6.22 (a) $C = \lim_{u \to \infty} e^{\kappa u} \psi(u) = \lim_{u \to \infty}\left[\frac{1}{4}e^{(\kappa-1)u} + \frac{1}{12}e^{(\kappa-2)u}\right]$. This will be a finite, positive number only when $\kappa = 1$ (in which case $C = 1/4$), and $\frac{1}{1+\theta} = \psi(0) = \frac{1}{4} + \frac{1}{12} = \frac{1}{3}$ and the loading is $\theta = 2$.
(b) $\frac{1}{4} = C = \frac{\theta \mu}{E(Xe^{\kappa X}) - \mu(1+\theta)} = \frac{2\mu}{2-3\mu}$ and so $\mu = \frac{2}{11}$. Then, $E\left(e^{\kappa X}\right) = 1 + (1+\theta)\mu\kappa = 1 + 3\left(\frac{2}{11}\right)(1) = \frac{17}{11}$.

6.23 $\lim_{u \to \infty} e^{r_1 u} \psi(u) = \lim_{u \to \infty}\left[C_1 + C_2 e^{-(r_2-r_1)u}\right] = C_1 \le 1, \quad 1 - C_1 \ge 0.$
$\phi(0) = 1 - \psi(0) = 1 - C_1 - C_2. \; C_2 = 1 - C_1 - \phi(0) \ge -\phi(0).$

6.24 (a) $X \sim \text{Gamma}(0.5, 1000)$, $\mu = 500$, $M_X(t) = E(e^{tX}) = (1-1000t)^{-1/2}$ and κ is the smallest root of $1 + (1+\theta)\mu\kappa = E(e^{\kappa X})$ which implies $1 + 750\kappa = (1 - 1000\kappa)^{-1/2}$. Then $(1 + 750\kappa)^2(1 - 1000\kappa) = 1$.

Let $r = 1000\kappa$. The equation becomes $r(9r^2 + 15r - 8) = 0$. The roots are 0 and $\frac{-15+\sqrt{513}}{18}$ and $\frac{-15+\sqrt{513}}{18}$. Hence $r = 0.425$ and $\kappa = 0.000425$. $E(Xe^{\kappa X}) = \frac{d}{d\kappa}E(e^{\kappa X}) = 500(1 - 1000\kappa)^{-3/2}$.

$$C = \frac{\mu\theta}{E(Xe^{\kappa X}) - \mu(1 + \theta)} = \frac{250}{500(1 - 1000\kappa)^{-3/2} - 750}$$
$$= [2(1 - 1000\kappa)^{-3/2} - 3]^{-1} = 0.630$$

and

$$\psi(u) \sim 0.630e^{-0.000425u}, \quad u \to \infty.$$

(b) $X \sim$ Uniform $(0, 100)$, $\mu = 50$, $M_X(t) = \frac{e^{100t}-1}{100t}$. κ is the smallest root of $1 + (1 + \theta)\mu\kappa = E(e^{\kappa X})$ which implies $1 + 75\kappa = \frac{e^{100\kappa}-1}{100\kappa}$ and so $e^{100\kappa} = 1 + 100\kappa + 7500\kappa^2$. Let $r = 100\kappa$. Then $e^r = 1 + r + \frac{3}{4}r^2$ and so $r = 1.10923$.

$$E(Xe^{\kappa X}) = \frac{d}{d\kappa}E(e^{\kappa X}) = 100\frac{re^r - (e^r - 1)}{r^2} = 108.19$$
$$C = \frac{\mu\theta}{E(Xe^{\kappa X}) - \mu(1 + \theta)} = \frac{25}{108.19 - 75} = 0.7535$$
$$\psi(u) \sim 0.7535e^{-0.0110923u}, \quad u \to \infty.$$

(c) $X \sim$ Geometric$(\beta = 9)$ and so $\mu = 9$ and

$$M_X(t) = P_X(e^t) = [1 - 9(e^t - 1)]^{-1}.$$

Then

$$E(Xe^{\kappa X}) = \frac{d}{d\kappa}[1 - 9(e^\kappa - 1)]^{-1} = 9[1 - 9(e^\kappa - 1)]^{-2}e^\kappa.$$

κ is the smallest positive root of $1 + (1 + \theta)\mu\kappa = E(e^{\kappa X})$. Hence $1 + 13.5\kappa = [1 - 9(e^\kappa - 1)]^{-1}$ which gives $\kappa = 0.0351$. Then

$$C = \frac{\mu\theta}{E(Xe^{\kappa X}) - \mu(1 + \theta)} = \frac{4.5}{20.25 - 13.5} = 0.667$$
$$\psi(u) \sim 0.667e^{-0.0351u}, \quad u \to \infty.$$

6.25 $\phi'(u) = e^{-\kappa u}m(u)$. We have

$$\psi(u) \sim Ce^{-\kappa u}, \quad u \to \infty \text{ which implies } \psi'(u) \sim -C\kappa e^{-\kappa u}, \quad u \to \infty.$$

Thus $\lim_{u \to \infty} e^{\kappa u}\psi'(u) = -C\kappa$. Now

$$\lim_{u \to \infty} m(u) = \lim_{u \to \infty} e^{\kappa u}\phi'(u) = \lim_{u \to \infty} e^{\kappa u}[-\psi'(u)]$$
$$= -\lim_{u \to \infty} e^{\kappa u}\psi'(u) = C\kappa.$$

Alternatively, use the renewal theorem. If $a(x) = b(x) + \int_0^x a(t-y)g(y)dy$, $x > 0$, then $\lim_{x \to \infty} a(x) = \dfrac{\int_0^\infty b(y)dy}{\int_0^\infty yg(y)dy}$. Now

$$m(u) = \frac{\theta}{(1+\theta)^2} e^{\kappa u} f_e(u) + \frac{1}{1+\theta} \int_0^u m(u-x)e^{\kappa x} f_e(x)dx.$$

We want to show

$$\lim_{u \to \infty} m(u) = \frac{\frac{\theta}{(1+\theta)^2} \int_0^\infty e^{\kappa y} f_e(y)dy}{\frac{1}{1+\theta} \int_0^\infty y e^{\kappa y} f_e(y)dy} = C\kappa.$$

From (6.8)

$$\frac{\theta}{(1+\theta)^2} \int_0^\infty e^{\kappa y} f_e(y)dy = \frac{\theta}{(1+\theta)^2}(1+\theta) = \frac{\theta}{1+\theta}$$

and so $\int_0^\infty e^{\kappa y} f_e(y)dy = 1 + \theta$. Also,

$$\begin{aligned}
\frac{1}{1+\theta} \int_0^\infty y e^{\kappa y} f_e(y)dy &= \frac{1}{1+\theta} \int_0^\infty \frac{1}{\kappa} y f_e(y) de^{\kappa y} \\
&= \frac{1}{\kappa(1+\theta)} \left[\left. \int_0^\infty y e^{\kappa y} f_e(y) \right|_0^\infty - \int_0^\infty e^{\kappa y} f_e(y)dy \right. \\
&\quad \left. + \frac{1}{\mu} \int_0^\infty y e^{\kappa y} dF(y) \right].
\end{aligned} \tag{5.22}$$

Since

$$\begin{aligned}
0 &\le y e^{\kappa y} f_e(y) = \frac{y e^{\kappa y}}{\mu} \int_y^\infty dF(x) \le \frac{1}{\mu} \int_y^\infty x e^{\kappa x} dF(x) \\
&\le \frac{1}{\mu} \int_0^\infty x e^{\kappa x} dF(x) = \frac{1}{\mu} E\left(X e^{\kappa X}\right),
\end{aligned}$$

$$E\left(X e^{\kappa X}\right) < \infty \Rightarrow \lim_{y \to \infty} \int_y^\infty x e^{\kappa x} dF(x) = 0 \Rightarrow \lim_{y \to \infty} y e^{\kappa y} f_e(y) = 0.$$

Then (5.22) implies

$$\frac{1}{\kappa(1+\theta)} \left[0 - 1 + \theta) + \frac{1}{\mu} E\left(X e^{\kappa X}\right) \right] = \frac{1}{\kappa \mu(1+\theta)} \left[E\left(X e^{\kappa X}\right) - \mu(1+\theta) \right]$$

and so

$$\begin{aligned}
\lim_{u \to \infty} m(u) &= \frac{\frac{\theta}{1+\theta}}{\frac{1}{\kappa \mu(1+\theta)} \left[E\left(X e^{\kappa X}\right) - \mu(1+\theta) \right]} \\
&= \frac{\theta \mu \kappa}{E\left(X e^{\kappa X}\right) - \mu(1+\theta)} = C\kappa
\end{aligned}$$

where $C = \frac{\theta \mu}{E(Xe^{\kappa X}) - \mu(1+\theta)}$.

6.26 (a) From (6.15)

$$\int_0^t \left[\frac{\partial}{\partial u} G(u, y) \right] du = \frac{\lambda}{c} \int_0^t G(u, y) du - \frac{\lambda}{c} \int_0^t \int_0^u G(u - x, y) dF(x) du$$

$$- \frac{\lambda}{c} \int_0^t [F(u + y) - F(u)] du.$$

Thus, by the fundamental theorem of calculus, and reversing the order of integration in the double integral,

$$G(t, y) - G(0, y) = \frac{\lambda}{c} \int_0^t G(u, y) du - \frac{\lambda}{c} \int_0^t \left[\int_x^t G(u - x, y) du \right] dF(x)$$

$$- \frac{\lambda}{c} \int_0^t [F(u + y) - F(u)] du .$$

Using (6.16) and changing the variable of integration from u to $v = u - x$ in the inner integral of the double integral results in

$$G(t, y) = \frac{\lambda}{c} \int_0^t G(u, y) du - \frac{\lambda}{c} \int_0^t \left[\int_0^{t-x} G(v, y) dv \right] dF(x)$$

$$+ \frac{\lambda}{c} \int_0^y [1 - F(x)] dx + \frac{\lambda}{c} \int_0^t [F(u + y) - F(u)] du.$$

For notational convenience, let $\Lambda(x, y) = \int_0^x G(v, y) dv$. Then

$$G(t, y) = \frac{\lambda}{c} \Lambda(t, y) - \frac{\lambda}{c} \int_0^t \Lambda(t - x, y) dF(x) + \frac{\lambda}{c} \int_0^y [1 - F(x)] dx$$

$$- \frac{\lambda}{c} \int_0^t [1 - F(u)] du + \frac{\lambda}{c} \int_0^t [1 - F(u + y)] du.$$

Then, integration by parts on the second integral on the right hand side and changing the variable of integration from u to $x = u + y$ on the first integral on the right hand side gives

$$G(t, y) = \frac{\lambda}{c} \Lambda(t, y) - \frac{\lambda}{c} \left[\Lambda(t - x, y) F(x) \big|_0^t + \int_0^t G(t - x, y) F(x) dx \right]$$

$$+ \frac{\lambda}{c} \int_0^y [1 - F(x)] dx - \frac{\lambda}{c} \int_0^t [1 - F(u)] du$$

$$+ \frac{\lambda}{c} \int_y^{y+t} [1 - F(x)] dx.$$

Thus,

$$G(t,y) = \frac{\lambda}{c}\Lambda(t,y) - \frac{\lambda}{c}\left[0 - 0 + \int_0^t G(t-x,y)F(x)dx\right]$$
$$+ \frac{\lambda}{c}\int_0^{y+t}[1-F(x)]dx - \frac{\lambda}{c}\int_0^t[1-F(u)]du.$$

Changing the variable of integration from u to x in the last integral yields, with $\Lambda(t,y) = \int_0^t G(t-x,y)dx$

$$G(t,y) = \frac{\lambda}{c}\int_0^t G(t-x,y)dx - \frac{\lambda}{c}\int_0^t G(t-x,y)F(x)dx + \frac{\lambda}{c}\int_t^{y+t}[1-F(x)]dx.$$

Finally, combining the first two integrals,

$$G(t,y) = \frac{\lambda}{c}\int_0^t G(t-x,y)[1-F(x)]dx + \frac{\lambda}{c}\int_t^{y+t}[1-F(x)]dx.$$

(b) From (a) we have the defective renewal equation

$$G(t,y) = \frac{1}{1+\theta}\int_0^t G(t-x,y)f_e(x)dx + \frac{1}{1+\theta}\int_t^{y+t} f_e(x)dx.$$

Multiplication of both sides by $e^{\kappa t}$ yields

$$G_*(t,y) = \int_0^t G_*(t-x,y)g(x)dx + \frac{\exp\left[\kappa t\int_t^{y+t}f_e(x)dx\right]}{1+\theta}$$

where $G_*(t,y) = e^{\kappa t}G(t,y)$, and $g(x) = e^{\kappa x}f_e(x)/(1+\theta)$ is a probability density function by (6.8). Thus, by the renewal theorem, (6.45) and (6.47):

$$\lim_{u\to\infty} G_*(u,y) = \frac{\int_0^\infty e^{\kappa t}\int_t^{y+t} f_e(x)dxdt}{(1+\theta)\int_0^\infty xg(x)dx}.$$

But, as shown in the proof of Theorem 6.7,

$$\int_0^\infty xg(x)dx = \frac{E(Xe^{\kappa X}) - \mu(1+\theta)}{\mu\kappa(1+\theta)}$$

and so

$$\lim_{u\to\infty} e^{\kappa u}G(u,y) = \frac{\frac{1}{\mu(1+\theta)}\int_0^\infty e^{\kappa t}\int_t^{y+t}[1-F(x)]dxdt}{\frac{1}{\mu\kappa(1+\theta)}[E(Xe^{\kappa X}) - \mu(1+\theta)]}$$

and the result follows.

5.6 SECTION 6.8

6.27 By Example 6.11, $\psi(u) = \frac{1}{1+\theta} \exp\left[-\frac{\theta u}{\mu(1+\theta)}\right]$.

$$\psi(1,000) = \frac{1}{1.1} \exp\left[-\frac{0.1(1000)}{100(1.1)}\right] = \frac{1}{1.1} \exp\left(-\frac{1}{1.1}\right) = 0.36626 \quad \text{(exact)}.$$

$$
\begin{aligned}
F_X(u) &= 1 - \exp(-u/100), E(X) = 100 \\
f_e(u) &= \frac{1 - F_x(u)}{E(X)} = \frac{1}{100} e^{-\frac{u}{100}}
\end{aligned}
$$

and so $F_e(u) = 1 - \exp(-u/100)$. Discretize $F_e(u)$ with a span of 50:

$$
\begin{aligned}
f_X(50m) &= \Pr\left[(2m-1)25 < u \le (2m+1)25\right] \\
&= \exp\left[-\frac{(2m-1)}{100} 25\right] - \exp\left[-\frac{(2m+1)}{100} 25\right] \\
&= \exp\left(-\frac{2m-1}{4}\right) - \exp\left(-\frac{2m+1}{4}\right), \quad m = 1, 2, \dots.
\end{aligned}
$$

$K \sim$ geometric with $p_k = \Pr(K = k) = \frac{\theta}{1+\theta}\left(\frac{1}{1+\theta}\right)^k$ and $p_k = \frac{1}{1+\theta} p_{k-1}$, $k = 1, 2, \dots$ Using

$$f_S(x) = \frac{1}{1 - a f_X(0)} \sum_{i=1}^{x} \left(a + b\frac{i}{x}\right) f_X(i) f_S(x - i)$$

with $a = \frac{1}{1+\theta}$, $b = 0$, $f_X(0) = 1 - \exp(-1/4)$ and

$$f_S(0) = \sum_{k=0}^{\infty} \frac{\theta}{1+\theta}\left(\frac{1}{1+\theta}\right)^k [f_X(0)]^k = \frac{\theta}{1+\theta - f_X(0)} = \frac{.1}{.1 + e^{-1/4}}$$

Calculations appear in Table 5.17.
Using the method of Section 4.9 we have

$$\phi(1,000) = \sum_{i=0}^{19} f_S(50 \times i) + 0.5 f_S(1000) = 0.637152.$$

Then, $\psi(1,000) = 1 - \phi(1,000) = 0.362858$. Using a span of 1 changes the result to 0.366263 which agrees with the analytical result to 5 decimal places.

6.28 By Example 6.13, $\psi(u) = \frac{2}{5} e^{-\frac{u}{2\beta}} - \frac{1}{15} e^{-\frac{4u}{3\beta}}$ and so

$$\psi(200) = \frac{2}{5} e^{-\frac{200}{100}} - \frac{1}{15} e^{-\frac{800}{150}} = \frac{2}{5} e^{-2} - \frac{1}{15} e^{-\frac{16}{3}} = 0.053812 \quad \text{(exact)}.$$

Table 5.17 Calculations for Exercise 6.27

x	$f_S(x)$	x	$f_S(x)$
0	0.113791	550	0.025097
50	0.039679	600	0.023973
100	0.037902	650	0.022900
150	0.036205	700	0.021875
200	0.034584	750	0.020895
250	0.033036	800	0.019960
300	0.031556	850	0.019066
350	0.030144	900	0.018212
400	0.028794	950	0.017397
450	0.027505	1,000	0.016618
500	0.026273		

$$f_X(x) = \beta^{-2}xe^{-x/\beta} = 50^{-2}xe^{-x/50}$$
$$E(X) = 2\beta = 100$$
$$F_X(x) = 1 - \frac{x}{50}e^{-x/50} - e^{-x/50}$$
$$f_e(u) = \frac{1 - F_X(u)}{E(X)} = \frac{1}{5,000}ue^{-u/50} + \frac{1}{100}e^{-u/50}$$
$$F_e(u) = 1 - \frac{1}{100}ue^{-u/50} - e^{-u/50}.$$

Although not relevant, this distribution is a mixture of gamma(2, 50) and gamma(1, 50) with probability one-half assigned to each. Discretize $F_e(u)$ with a span of 25.

$$f_X(25m) = \Pr\left[(2m-1)12.5 < Y \le (2m+1)12.5\right]$$
$$= \frac{1}{100}[(2m-1)12.5 - (2m+1)12.5]$$
$$+ \left[e^{-(2m-1)/4} - e^{-(2m+1)/4}\right], \quad m = 1, 2, \ldots.$$

Using

$$f_X(x) = \frac{1}{1 - af_x(0)}\sum_{i=1}^{x}\left(a + b\frac{i}{x}\right)f_X(i)f_S(x-i)$$

where $a = \frac{1}{1+\theta} = \frac{1}{3}$, $b = 0$, $f_X(0) = 1 - \frac{125}{100}\exp(-12.5/50) - \exp(-12.5/50)$ and

$$f_S(0) = \frac{\theta}{1 + \theta - f_X(0)} = \frac{2}{2 + \frac{12.5}{100}e^{-1/4} + e^{-1/4}}.$$

Therefore,

$$f_S(x) = \frac{1}{3 - f_X(0)}\sum_{i=0}^{x}f_X(i)f_S(x-i)$$

We then have the calculations given in Table 5.18.

Table 5.18   Calculations for Exercise 6.28

x	$f_S(x)$
0	0.695374
25	0.054797
50	0.048788
75	0.041135
100	0.033649
125	0.027036
150	0.021483
175	0.016952
200	0.013317

Using the method of Section 4.9 we have

$$\phi(200) = \sum_{m=0}^{7} f_S(25m) + 0.5 f_S(200) = 0.9458725$$

and $\psi(200) = 1 - \phi(200) = 0.0541275$.

6.29 (a) First, change the index of summation from k to j, yielding

$$F'(x) = \sum_{j=1}^{r} q_j \frac{\beta^{-j} x^{j-1} e^{-x/\beta}}{(j-1)!} .$$

Then, from Example 4.7,

$$
\begin{aligned}
F(x) &= \int_0^x F'(y) dy = \sum_{j=1}^{r} q_j \int_0^x \frac{\beta^{-j} y^{j-1} e^{-y/\beta}}{(j-1)!} dy = \sum_{j=1}^{r} q_j \Gamma(j; x/\beta) \\
&= \sum_{j=1}^{r} q_j \left\{ 1 - \sum_{k=1}^{j} \frac{(x/\beta)^{k-1} e^{-x/\beta}}{(k-1)!} \right\} = 1 - \sum_{j=1}^{r} \sum_{k=1}^{j} q_j \frac{(x/\beta)^{k-1} e^{-x/\beta}}{(k-1)!} \\
&= 1 - \sum_{k=1}^{r} \sum_{j=k}^{r} q_j \frac{(x/\beta)^{k-1} e^{-x/\beta}}{(k-1)!} = 1 - \sum_{k=1}^{r} \frac{(x/\beta)^{k-1} e^{-x/\beta}}{(k-1)!} \sum_{j=k}^{r} q_j .
\end{aligned}
$$

Also,

$$\mu = \int_0^\infty x F'(x) dx = \sum_{j=1}^{r} q_j \int_0^\infty x \cdot \frac{\beta^{-j} x^{j-1} e^{-x/\beta}}{(j-1)!} dx = \sum_{j=1}^{r} q_j (\beta j) = \beta \sum_{j=1}^{r} j q_j .$$

Thus,

$$f_e(x) = \frac{1 - F(x)}{\mu} = \frac{\sum\limits_{k=1}^{r} \frac{(x/\beta)^{k-1} e^{-x/\beta}}{(k-1)!} \sum\limits_{j=k}^{r} q_j}{\beta \sum\limits_{j=1}^{r} j q_j} = \sum\limits_{k=1}^{r} q_k^* \frac{\beta^{-k} x^{k-1} e^{-x/\beta}}{(k-1)!}$$

It remains to show that $\sum\limits_{k=1}^{r} q_k^* = 1$. By interchanging the order of summation,

$$\sum_{k=1}^{r} \sum_{j=k}^{r} q_j = \sum_{j=1}^{r} \sum_{k=1}^{j} q_j = \sum_{j=1}^{r} j q_j.$$

Division of both sides by $\sum\limits_{j=1}^{r} j q_j$ gives the result. Thus, $f_e(x)$ is of the same form as $F'(x)$, but with the mixing weights $\{q_j; j = 1, 2, \ldots, r\}$ replaced by $\{q_j^*; j = 1, 2, \ldots, r\}$.
(b) From Exercise 4.35(a),

$$\int_0^\infty e^{zx} f_e(x) dx = Q^*\{(1 - \beta z)^{-1}\}.$$

Thus, from Exercise 4.35(b), the maximum aggregate loss L has moment generating function

$$E(e^{zL}) = \sum_{k=0}^{\infty} \frac{\theta}{1+\theta} \left(\frac{1}{1+\theta}\right)^k (Q^*\{(1 - \beta z)^{-1}\})^k$$

$$= \frac{\theta}{\theta + 1 - Q^*\{(1 - \beta z)^{-1}\}}.$$

That is, $E(e^{zL}) = C\{(1 - \beta z)^{-1}\}$ after divsion of the numerator and denominator by θ. Clearly, $C(z) = \sum\limits_{n=0}^{\infty} c_n z^n$ is the pgf of a compound geometric distribution with $a = (1 + \theta)^{-1}$ and $b = 0$, and secondary "claim size" pgf $Q^*(z)$. Thus, by Theorem 3.3, the probabilities $\{c_n; n = 0, 1, 2, \ldots\}$ may be computed recursively by

$$c_k = \frac{1}{1+\theta} \sum_{j=1}^{k} q_j^* c_{k-j}; \quad k = 1, 2, \ldots$$

beginning with $c_0 = \theta(1 + \theta)^{-1}$. Then from Exercise 4.35,

$$\psi(u) = \sum_{n=1}^{\infty} c_n \sum_{j=0}^{n-1} \frac{(u/\beta)^j e^{-u/\beta}}{j!}, \quad u \geq 0.$$

(c) From (b), interchanging the order of summation,

$$\psi(u) = e^{-u/\beta} \sum_{j=0}^{\infty} \sum_{n=j+1}^{\infty} \frac{(u(\beta)^j}{j!} c_n = e^{-u/\beta} \sum_{j=0}^{\infty} \bar{C}_j \frac{(u/\beta)^j}{j!}.$$

Clearly, $\bar{C}_0 = 1 - c_0 = (1 + \theta)^{-1}$. By summing the recursion in (b),

$$\sum_{k=n+1}^{\infty} c_k = \frac{1}{1+\theta} \sum_{k=n+1}^{\infty} \sum_{j=1}^{k} q_j^* c_{k-j}.$$

Interchanging the order of summation yields

$$\bar{C}_n = \frac{1}{1+\theta} \sum_{j=1}^{n} q_j^* \sum_{k=n+1}^{\infty} c_{k-j} + \frac{1}{1+\theta} \sum_{j=n+1}^{\infty} q_j^* \sum_{k=j}^{\infty} c_{k-j}.$$

But $\bar{C}_{n-j} = \sum_{k=n+1}^{\infty} c_{k-j}$ and $1 = \sum_{k=j}^{\infty} c_{k-j}$, yielding

$$\bar{C}_n = \frac{1}{1+\theta} \sum_{j=1}^{n} q_j^* \bar{C}_{n-j} + \frac{1}{1+\theta} \sum_{j=n+1}^{\infty} q_j^*.$$

(d) Consider

$$\int_0^{\infty} e^{zx} F'(x) dx = E(e^{zX}).$$

By integration by parts,

$$\int_0^{\infty} e^{zx} F'(x) dx = -e^{zx} \{1 - F(x)\}|_0^{\infty} + z \int_0^{\infty} e^{zx} \{1 - F(x)\} dx$$

$$= 1 + \mu z \int_0^{\infty} e^{zx} f_e(x) dx$$

since $f_e(x) = \{1 - F(x)\}/\mu$. Also

$$0 \le \lim_{x \to \infty} e^{zx} \{1 - F(x)\} = \lim_{x \to \infty} e^{zx} \int_x^{\infty} F'(y) dy \le \lim_{x \to \infty} \int_x^{\infty} e^{\kappa y} F'(y) dy = 0$$

if $E(e^{zx}) < \infty$. That is, $\lim_{x \to \infty} e^{zx} [1 - F(x)] = 0$. In other words, the moment generating functions of X and Y are related by

$$E(e^{zx}) = 1 + \mu z E(e^{zY}).$$

Differentiating with respect to z yields

$$E(Xe^{zX}) = \mu E(e^{zY}) + \mu z E(Ye^{zY}).$$

Thus, with $z = \kappa$,

$$
\begin{aligned}
E(Xe^{\kappa X}) &= \mu E(e^{\kappa Y}) + \mu \kappa E(Ye^{\kappa Y}) \\
&= \mu(1 + \theta) + \mu \kappa E(Ye^{\kappa Y})
\end{aligned}
$$

using (6.8). Thus, from the above and (6.44),

$$C = \frac{\theta \mu}{\mu \kappa E(Ye^{\kappa Y})} = \frac{\theta}{\kappa E(Ye^{\kappa Y})}.$$

In this case, from (b)

$$E(e^{zY}) = \int_0^\infty e^{zx} f_e(x)dx = Q^*\{(1 - \beta z)^{-1}\} = \sum_{k=1}^r q_k^*(1 - \beta z)^{-k}.$$

Differentiating with respect to z gives

$$E(Ye^{zY}) = \beta \sum_{k=1}^r kq_k^*(1 - \beta z)^{-k-1} = \beta \sum_{j=1}^r jq_j^*(1 - \beta z)^{-j-1}.$$

Thus, with z replaced by κ,

$$C = \frac{\theta}{\kappa E(Ye^{\kappa Y})} = \frac{\theta}{\kappa \beta \sum_{j=1}^r jq_j^*(1 - \beta \kappa)^{-j-1}}$$

and Cramér's asymptotic ruin formula gives $\psi(u) \sim Ce^{-\kappa u}, u \to \infty$. Finally, from (6.8), $\kappa > 0$ satisfies

$$
\begin{aligned}
1 + \theta &= E \int_0^\infty e^{\kappa y} f_e(y)dy = E(e^{\kappa Y}) \\
&= Q^*\{(1 - \beta\kappa)^{-1}\} = \sum_{j=1}^r q_j^*(1 - \beta\kappa)^{-j}.
\end{aligned}
$$

Texts and References Section

AGRESTI · An Introduction to Categorical Data Analysis

ANDERSON · An Introduction to Multivariate Statistical Analysis, *Second Edition*

ANDERSON and LOYNES · The Teaching of Practical Statistics

ARMITAGE and COLTON · Encyclopedia of Biostatistics: Volumes 1 to 6 with Index

BARTOSZYNSKI and NIEWIADOMSKA-BUGAJ · Probability and Statistical Inference

BERRY, CHALONER, and GEWEKE · Bayesian Analysis in Statistics and Econometrics: Essays in Honor of Arnold Zellner

BHATTACHARYA and JOHNSON · Statistical Concepts and Methods

BILLINGSLEY · Probability and Measure, *Second Edition*

BOX · R. A. Fisher, the Life of a Scientist

BOX, HUNTER, and HUNTER · Statistics for Experimenters: An Introduction to Design, Data Analysis, and Model Building

BOX and LUCEÑO · Statistical Control by Monitoring and Feedback Adjustment

BROWN and HOLLANDER · Statistics: A Biomedical Introduction

CHATTERJEE and PRICE · Regression Analysis by Example, *Second Edition*

COOK and WEISBERG · An Introduction to Regression Graphics

COX · A Handbook of Introductory Statistical Methods

DILLON and GOLDSTEIN · Multivariate Analysis: Methods and Applications

DODGE and ROMIG · Sampling Inspection Tables, *Second Edition*

DRAPER and SMITH · Applied Regression Analysis, *Third Edition*

DUDEWICZ and MISHRA · Modern Mathematical Statistics

DUNN · Basic Statistics: A Primer for the Biomedical Sciences, *Second Edition*

FISHER and VAN BELLE · Biostatistics: A Methodology for the Health Sciences

FREEMAN and SMITH · Aspects of Uncertainty: A Tribute to D. V. Lindley

GROSS and HARRIS · Fundamentals of Queueing Theory, *Third Edition*

HALD · A History of Probability and Statistics and their Applications Before 1750

HALD · A History of Mathematical Statistics from 1750 to 1930

HELLER · MACSYMA for Statisticians

HOEL · Introduction to Mathematical Statistics, *Fifth Edition*

JOHNSON and BALAKRISHNAN · Advances in the Theory and Practice of Statistics: A Volume in Honor of Samuel Kotz

JOHNSON and KOTZ (editors) · Leading Personalities in Statistical Sciences: From the Seventeenth Century to the Present

JUDGE, GRIFFITHS, HILL, LÜTKEPOHL, and LEE · The Theory and Practice of Econometrics, *Second Edition*

KHURI · Advanced Calculus with Applications in Statistics

KOTZ and JOHNSON (editors) · Encyclopedia of Statistical Sciences: Volumes 1 to 9 wtih Index

KOTZ and JOHNSON (editors) · Encyclopedia of Statistical Sciences: Supplement Volume

KOTZ, REED, and BANKS (editors) · Encyclopedia of Statistical Sciences: Update Volume 1

KOTZ, REED, and BANKS (editors) · Encyclopedia of Statistical Sciences: Update Volume 2

LAMPERTI · Probability: A Survey of the Mathematical Theory, *Second Edition*

LARSON · Introduction to Probability Theory and Statistical Inference, *Third Edition*

LE · Applied Survival Analysis

MALLOWS · Design, Data, and Analysis by Some Friends of Cuthbert Daniel

MARDIA · The Art of Statistical Science: A Tribute to G. S. Watson

MASON, GUNST, and HESS · Statistical Design and Analysis of Experiments with Applications to Engineering and Science

MURRAY · X-STAT 2.0 Statistical Experimentation, Design Data Analysis, and Nonlinear Optimization

*Now available in a lower priced paperback edition in the Wiley Classics Library.

Applied Probability and Statistics (Continued)

McLACHLAN and KRISHNAN · The EM Algorithm and Extensions
McLACHLAN · Discriminant Analysis and Statistical Pattern Recognition
McNEIL · Epidemiological Research Methods
MILLER · Survival Analysis
MONTGOMERY and PECK · Introduction to Linear Regression Analysis, *Second Edition*
MYERS and MONTGOMERY · Response Surface Methodology: Process and Product in Optimization Using Designed Experiments
NELSON · Accelerated Testing, Statistical Models, Test Plans, and Data Analyses
NELSON · Applied Life Data Analysis
OCHI · Applied Probability and Stochastic Processes in Engineering and Physical Sciences
OKABE, BOOTS, and SUGIHARA · Spatial Tesselations: Concepts and Applications of Voronoi Diagrams
PANKRATZ · Forecasting with Dynamic Regression Models
PANKRATZ · Forecasting with Univariate Box-Jenkins Models: Concepts and Cases
PIANTADOSI · Clinical Trials: A Methodologic Perspective
PORT · Theoretical Probability for Applications
PUTERMAN · Markov Decision Processes: Discrete Stochastic Dynamic Programming
RACHEV · Probability Metrics and the Stability of Stochastic Models
RÉNYI · A Diary on Information Theory
RIPLEY · Spatial Statistics
RIPLEY · Stochastic Simulation
ROUSSEEUW and LEROY · Robust Regression and Outlier Detection
RUBIN · Multiple Imputation for Nonresponse in Surveys
RUBINSTEIN · Simulation and the Monte Carlo Method
RUBINSTEIN, MELAMED, and SHAPIRO · Modern Simulation and Modeling
RYAN · Statistical Methods for Quality Improvement
SCHUSS · Theory and Applications of Stochastic Differential Equations
SCOTT · Multivariate Density Estimation: Theory, Practice, and Visualization
*SEARLE · Linear Models
SEARLE · Linear Models for Unbalanced Data
SEARLE, CASELLA, and McCULLOCH · Variance Components
STOYAN, KENDALL, and MECKE · Stochastic Geometry and Its Applications, *Second Edition*
STOYAN and STOYAN · Fractals, Random Shapes and Point Fields: Methods of Geometrical Statistics
THOMPSON · Empirical Model Building
THOMPSON · Sampling
TIJMS · Stochastic Modeling and Analysis: A Computational Approach
TIJMS · Stochastic Models: An Algorithmic Approach
TITTERINGTON, SMITH, and MAKOV · Statistical Analysis of Finite Mixture Distributions
UPTON and FINGLETON · Spatial Data Analysis by Example, Volume 1: Point Pattern and Quantitative Data
UPTON and FINGLETON · Spatial Data Analysis by Example, Volume II: Categorical and Directional Data
VAN RIJCKEVORSEL and DE LEEUW · Component and Correspondence Analysis
WEISBERG · Applied Linear Regression, *Second Edition*
WESTFALL and YOUNG · Resampling-Based Multiple Testing: Examples and Methods for *p*-Value Adjustment
WHITTLE · Systems in Stochastic Equilibrium
WOODING · Planning Pharmaceutical Clinical Trials: Basic Statistical Principles
WOOLSON · Statistical Methods for the Analysis of Biomedical Data
*ZELLNER · An Introduction to Bayesian Inference in Econometrics

*Now available in a lower priced paperback edition in the Wiley Classics Library.

*Now available in a lower priced paperback edition in the Wiley Classics Library.

*Now available in a lower priced paperback edition in the Wiley Classics Library.

*Now available in a lower priced paperback edition in the Wiley Classics Library.

WILEY SERIES IN PROBABILITY AND STATISTICS

ESTABLISHED BY WALTER A. SHEWHART AND SAMUEL S. WILKS

Editors

Vic Barnett, Ralph A. Bradley, Noel A. C. Cressie, Nicholas I. Fisher,
Iain M. Johnstone, J. B. Kadane, David G. Kendall, David W. Scott,
Bernard W. Silverman, Adrian F. M. Smith, Jozef L. Teugels,
Geoffrey S. Watson; J. Stuart Hunter, Emeritus

Probability and Statistics Section

*ANDERSON · The Statistical Analysis of Time Series
ARNOLD, BALAKRISHNAN, and NAGARAJA · A First Course in Order Statistics
BACCELLI, COHEN, OLSDER, and QUADRAT · Synchronization and Linearity:
   An Algebra for Discrete Event Systems
BASILEVSKY · Statistical Factor Analysis and Related Methods: Theory and
   Applications
BERNARDO and SMITH · Bayesian Statistical Concepts and Theory
BILLINGSLEY · Convergence of Probability Measures
BOROVKOV · Asymptotic Methods in Queuing Theory
BRANDT, FRANKEN, and LISEK · Stationary Stochastic Models
CAINES · Linear Stochastic Systems
CAIROLI and DALANG · Sequential Stochastic Optimization
CONSTANTINE · Combinatorial Theory and Statistical Design
COVER and THOMAS · Elements of Information Theory
CSÖRGŐ and HORVÁTH · Weighted Approximations in Probability Statistics
CSÖRGŐ and HORVÁTH · Limit Theorems in Change Point Analysis
DETTE and STUDDEN · The Theory of Canonical Moments with Applications in
   Statistics, Probability, and Analysis
*DOOB · Stochastic Processes
DRYDEN and MARDIA · Statistical Analysis of Shape
DUPUIS and ELLIS · A Weak Convergence Approach to the Theory of Large Deviations
ETHIER and KURTZ · Markov Processes: Characterization and Convergence
FELLER · An Introduction to Probability Theory and Its Applications, Volume 1,
   Third Edition, Revised; Volume II, *Second Edition*
FULLER · Introduction to Statistical Time Series, *Second Edition*
FULLER · Measurement Error Models
GELFAND and SMITH · Bayesian Computation
GHOSH, MUKHOPADHYAY, and SEN · Sequential Estimation
GIFI · Nonlinear Multivariate Analysis
GUTTORP · Statistical Inference for Branching Processes
HALL · Introduction to the Theory of Coverage Processes
HAMPEL · Robust Statistics: The Approach Based on Influence Functions
HANNAN and DEISTLER · The Statistical Theory of Linear Systems
HUBER · Robust Statistics
IMAN and CONOVER · A Modern Approach to Statistics
JUREK and MASON · Operator-Limit Distributions in Probability Theory
KASS and VOS · Geometrical Foundations of Asymptotic Inference
KAUFMAN and ROUSSEEUW · Finding Groups in Data: An Introduction to Cluster
   Analysis

*Now available in a lower priced paperback edition in the Wiley Classics Library.